Interdisciplinary Lively Application Projects (ILAPs)

This project was supported by the National Science Foundation (DUE 9455980). The comments and opinions in this report are those of the authors and not necessarily those of the National Science Foundation.

This volume is not endorsed by the Army or the Department of Defense. Opinions expressed by authors employed by the Department of Defense are those of the authors and not those of the Army or Department of Defense.

Library of Congress Catalog Card Number 97-70504
ISBN 0-88385-706-5
Printed in the United States of America
Current Printing (last digit):
10 9 8 7 6 5 4 3 2 1

Interdisciplinary Lively Application Projects (ILAPs)

David C. Arney, Editor
United States Military Academy

Published and distributed by
The Mathematical Association of America

CLASSROOM RESOURCE MATERIALS

This series provides supplementary material for students and their teachers—laboratory exercises, projects, historical information, textbooks with unusual approaches for presenting mathematical ideas, career information, and much more.

Proofs Without Words, by Roger Nelsen
A Radical Approach to Real Analysis, by David Bressoud
She Does Math! Real-Life Problems from Women on the Job, edited by Marla Parker
Learn from the Masters! edited by Frank Swetz, John Fauvel, Otto Bekken, Bengt Johansson, and Victor Katz
101 Careers in Mathematics, edited by Andrew Sterrett
Laboratory Experiences in Group Theory, by Ellen Maycock Parker
Interdisciplinary Lively Application Projects (ILAPs), edited by David C. Arney

These volumes can be ordered from:
MAA Service Center
P.O. Box 91112
Washington, DC 20090-1112
1-800-331-1MAA FAX: 1-301-206-9789

Preface

The United States Military Academy leads a consortium of 12 schools located throughout the country in a program to improve the educational culture through the enhancement of interdisciplinary cooperation and coordination. The consortium's program, entitled Project INTERMATH, is funded by a National Science Foundation (NSF) grant under the initiative "Mathematical Sciences and their Applications Throughout the Curriculum." The primary activity of Project INTERMATH is the development and use of Interdisciplinary Lively Application Projects (ILAPs). These small group projects are developed through the interdisciplinary cooperation of faculty from mathematics and partner disciplines. The consortium has begun the production of ILAPs, which have goals of welding mathematics with the concepts and principles of partner disciplines, promoting curricular reorganization by focusing on student growth in problem solving, and promoting faculty growth through the ILAP development processes.

From the student's perspective, ILAPs provide applications which motivate the need to develop mathematical concepts and skills, provide interest in future subjects that become accessible through further study and mastery of mathematics, and enable a broader, more interdisciplinary outlook at an earlier stage of development. From a faculty perspective, ILAPs are valuable tools to accomplish a variety of course goals. In accomplishing these goals, Project INTERMATH will create an infrastructure of schools and faculty that will have the potential for national educational reform.

This volume contains a considerable amount of classroom resource material in the form of project handouts that instructors can use for student homework assignments. These ILAPs are interdisciplinary in their nature, terminology, and notation. They usually require students to use scientific and quantitative reasoning, mathematical modeling, symbolic manipulation skills, and computational tools to solve and analyze scenarios, issues, and questions involving one or more disciplines. The student effort can involve individual or group work, and the final product can be presented in oral and written reports. The actual time spent by students (student-team) to solve and present an ILAP is difficult to predict. The ILAPs in this volume have taken 6-10 hours of student effort on the average. However, as with all large or open-ended problems, there will be a large variation in time spent by different student teams. Guidance from the faculty on the time allocated to work on ILAPs is important.

The material in this volume is formatted so that instructors can duplicate the student hand-out pages and provide them directly to their students. In addition, there are other project-related materials in this volume. Sample solutions, background material, instructor notes, and a student writing guide are included as needed for the projects. These can be used by the instructor for student handouts, after the student solution has been submitted, or as classroom discussion material.

The level of prerequisite skills for the eight projects presented in this volume range from freshman-level algebra, trigonometry, and precalculus; through calculus, elementary differential equations, and discrete mathematics; through intermediate differential equations; to advanced calculus and partial differential equations. The partner disciplines included in the projects are: mechanics, physics, chemistry, engineering, geography, topography, and exercise physiology. Instructors in almost any undergraduate applied mathematics course (precalculus, calculus, linear algebra, differential equations, discrete mathematics, mathematical modeling, advanced calculus, partial differential equations, numerical computing, etc.) could use portions of these projects in their courses. The projects are appropriate for both traditional and reformed courses, although, for many of the projects, computing technology is very helpful.

This volume also contains several supporting articles that describe uses for these projects and explanations of other components of Project INTERMATH. Other aspects of Project INTERMATH include promoting faculty communication; developing, studying, and testing curricula; and conducting faculty development workshops and conferences. At the first INTERMATH consortium workshop devoted to writing, interdisciplinary teams from eleven schools spent a weekend ILAPs at West Point in November 1996 writing ILAPs and planning for their use in courses.

The NSF has taken the initiative in interdisciplinary education and is currently supporting seven projects to make advances in our interdisciplinary culture. The MAA's Committee for the Undergraduate Program in Mathematics (CUPM) has established a special subcommittee, chaired by Frank Giordano, to study, guide, and promote progress in this area. I believe that faculty can effectively and efficiently use the material in this volume to accomplish the important goal of giving students opportunities to solve problems in an interdisciplinary setting.

Acknowledgments

By its very nature, Project INTERMATH is a cooperative effort. To produce a volume like this one, the editor needs a talented, cooperative, interdisciplinary team. Fortunately, many talented and cooperative people volunteered to be part of the team. I thank all the authors for their efforts in meeting deadlines and producing such outstanding products. Thanks to the editorial helpers Kathleen Snook, Jon Christensen, Joseph Myers, and Brian Winkel. Thanks to Michael Johnson, Dean Mengel, and Jon Christensen for their help with the graphics contained in this volume. Thanks to Lisa M. Arney for her outstanding photography. Thanks to the professionals at COMAP, who are partners and helpers in this project. Thanks to the directors of Project INTERMATH, Richard West, Donald Small, James Stith, Jack Grubbs, and Frank Giordano.

I thank the leadership of the United States Military Academy, especially the Academic Dean Fletcher M. Lamkin, for their vision of interdisciplinary and multidisciplinary education for cadets and for their support for this project. The cooperation and dedicated efforts of the members of the Mathematics-Science-Engineering (MSE) Committee at West Point have been a model for cultural change in education and my personal inspiration for work on this project.

Finally, thanks to the following organizations involved in supporting this effort and to their dedicated employees: National Science Foundation, which generously supports Project INTERMATH; and the Mathematics Association of America (MAA), publisher of this series on classroom resource materials. Special appreciation to the MAA staff members, Don Albers, Elaine Pedreira, and Beverly Ruedi, for being so friendly and helpful during the final stages of production. I sincerely thank the members of the MAA's Editorial Board for Classroom Resource Materials, chaired by Andrew Sterrett. Their support and professional editing were instrumental to the success of this writing project and to the publication of this classroom resource.

David C. Arney
Editor

Contents

Articles

Interdisciplinary Lively Application Project

Title: Getting Fit With Mathematics

Authors: Joseph D. Myers
Walter S. Barge
Todd A. Crowder
Kathleen G. Snook

Department of Mathematical Sciences and Department of Physical Education, United States Military Academy, West Point, New York

Editors: David C. Arney and Jon L. Christensen

Mathematics Classifications: Algebra, Pre-Calculus, Calculus

Disciplinary Classification: Exercise Physiology

Prerequisite Skills:

1. Graphing and Analyzing Graphs
2. Analyzing Linear and Nonlinear Functions
3. Composing Functions
4. Elementary Curve Fitting
5. Elementary Integration (Polynomials)
6. Elementary Numerical Integration

Physical Concepts Examined:

1. Oxygen Consumption During Exercise
2. Power During Exercise
3. Heart Rate During Exercise

Materials:

1. Problem Statement (2 Situations, 10 Parts); Student
2. Sample Solution (2 Situations, 10 Parts); Instructor
3. Notes for the Instructor

Computing Requirements:

1. Graphing Package and Spreadsheet are helpful tools for visualization of functions in this problem
2. Calculator or Computer to Perform Regression

With the emphasis on fitness these days, the Physical Education Department would like to institute a program that offers free exercise classes to students. In order to obtain funding to pay for instructors, the physical education department must make a presentation to the school's board of trustees explaining the program's benefits. To support the Physical Education Department's efforts, you have been asked to do some initial mathematical analysis on the benefits of exercise in terms of improved oxygen consumption and more efficient generation of the cellular fuel adenosine triphosphate (ATP).

The Physical Education Department provided you the following graphs depicting relationships between exercise duration, power exerted, oxygen consumption rate, and lactic acid release rate. The graphs given below were produced by averaging data collected during student fitness tests.

Graph 1. Power expended at any instant during the exercise is a function of time. This graph shows a typical laboratory treadmill test. Initially, the treadmill is horizontal and moving at an easy pace. Over time, the treadmill is gradually inclined and the treadmill belt speed is gradually increased (both at a constant rate of increase). The exerciser must use more and more power to "keep up with" and "stay on" the treadmill.

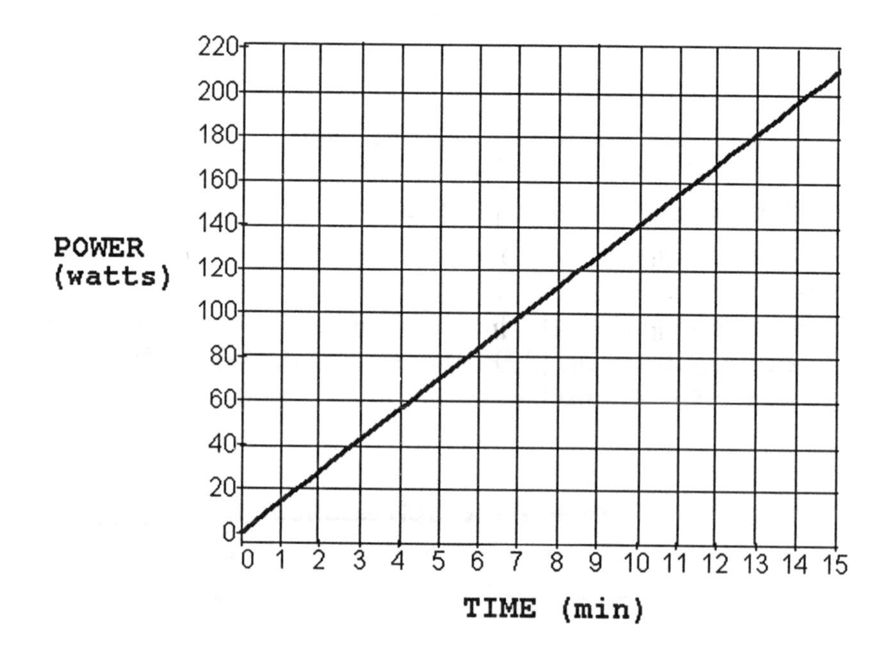

Graph 2. The rate at which oxygen is consumed during exercise is a function of power expended at that instant. Here the graph shows that, when power expended is low, the oxygen consumption rate is also low. As the treadmill test requires a greater expenditure of power, the oxygen consumption rate increases up to a maximum point for the individual being tested--the max VO_2 (volume of oxygen). The test is concluded when the runner reaches max VO_2 and can no longer "keep up with" or "stay on" the treadmill. This graph shows the average for the student fitness tests.

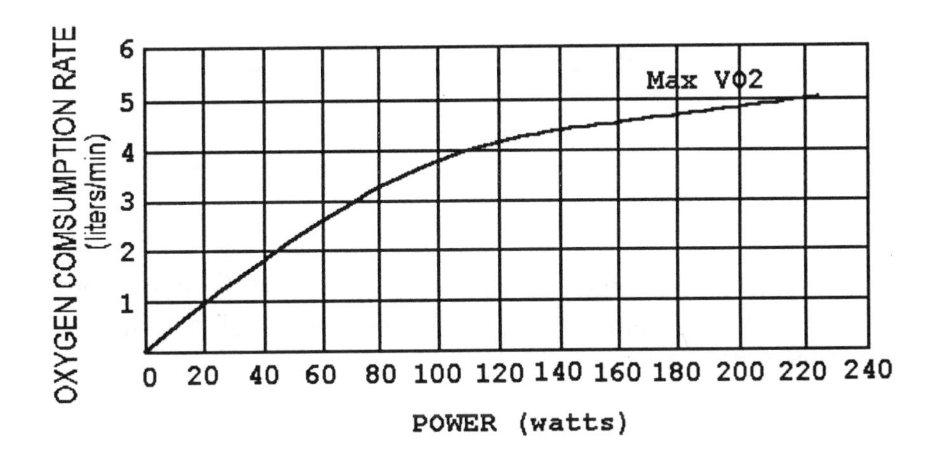

Graph 3. The following graph shows that lactic acid formation in the blood is a function of the rate of oxygen consumption. Some lactic acid is produced early in the test even when the oxygen consumption rate is low. Notice that, when the onset of blood lactic acid (OBLA) begins, the concentration of lactic acid in the blood increases dramatically. Again, this graph shows the average for the students in the fitness test.

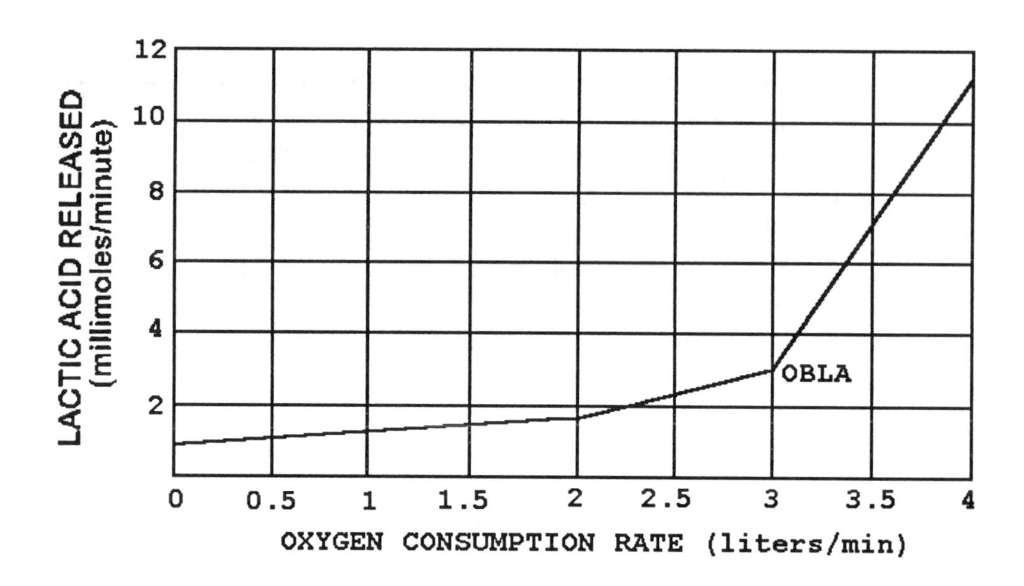

Background Information

In order for you to become familiar with physical fitness factors, the Physical Education Department provided you with the following background information.

The ability to sustain a high level of physical activity without undue fatigue depends on two factors: (1) oxygen delivery and (2) the capacity of specific muscle cells to generate the cellular fuel adenosine triphosphate, or ATP (McArdle, 1991, p. 223). The formation of ATP to be used for muscular energy begins when glucose molecules undergo a chemical transformation in a process known as glycolysis. When the body is subject to light exercise, even of a long duration, the ATP is produced through an efficient slow glycolysis. End products from the glycolysis (ADP) are easily removed from muscles at about the same rate that they are produced, and there is little accumulation. Under conditions of strenuous exercise, the demand for ATP can exceed the cell's ability to produce it efficiently. When this happens, the muscle cells resort to an inefficient fast glycolysis which releases end products faster than they can be removed from the muscles. The result is an accumulation of end products in the muscle fibers and blood stream. One of these end products is lactic acid. The fast glycolysis buys time for the muscles by rapidly producing ATP even if the oxygen supply is inadequate or the exercise too strenuous to produce ATP efficiently (McArdle, 1991, p. 125).

However, using fast glycolysis to meet muscular energy demands is only a temporary solution; the accumulation of lactic acid in the blood stream contributes to muscle fatigue which prevents continued physical activity. Effective aerobic conditioning is one way to enhance the capacity of specific muscle cells to generate ATP efficiently and thus delay muscle fatigue.

The capacity for oxygen (O_2) consumption is a fundamental measure of maximal aerobic power (McArdle, 1991, p. 211). The highest rate of oxygen consumption during a controlled fitness test is called max VO_2 (i.e., Value of O_2) and is measured in liters/minute of oxygen consumed. No one can operate for a long time at one's max VO_2 level, but there is a significant connection between a person's max VO_2 and the formation of lactic acid in the blood. For people of all levels of physical fitness, when engaging in strenuous physical activity there is a certain percentage of max VO_2 at which the production of blood lactic acid shows a near exponential increase. This point is called the Onset of Blood Lactic Acid (OBLA) (McArdle, 1991, p. 126). As stated above, endurance is influenced by the oxygen delivery rate, which is linked to max VO_2, and by the generation of ATP, which is linked to the point where OBLA begins.

Project References

McArdle, W.D., Katch, F.II and Katch, V.L. (1991). *Exercise physiology.* Malvern, PA: Lea and Febiger.

Sharkey, B.J. (1990). *Physiology of fitness.* Champaign, IL: Human Kinetics Books.

Situation 1. The Physical Education Department wants you to do an initial analysis of the relationship of activity and fitness to health. One of the components of fitness is aerobic fitness, which refers to the capacity to take in, transport, and utilize oxygen. As indicated in the background information, effective aerobic conditioning is one way to enhance the capacity of specific muscle cells to generate cellular fuel. This cellular fuel helps to delay muscle fatigue. In other words, the better your level of aerobic conditioning, the more energy and endurance you will have for physical activities.

Power and Oxygen Consumption

The capacity for oxygen consumption is a fundamental measure of maximal aerobic power. Additionally, when power expended in exercise is low, the rate of oxygen consumption is also low; when power expended is higher, the rate of oxygen consumption is also higher. Therefore, improving our level of aerobic conditioning results in being able to exercise at a higher power level, thereby increasing our oxygen consumption rate, which in turn benefits our cardiovascular and cardio-respiratory systems. Because aerobic activities such as walking, jogging, cycling, swimming, and aerobic dance involve so many important organs and systems (heart and circulation, respiratory, muscles), aerobic fitness is a good indication of health in general.

For the following requirements use **Graphs 1 - 3**.

Part 1.

A. Which of the graphs appear(s) to be linear? Explain your choice(s).

B. For each of the three graphs, develop a table of input and output values. Which of the graphs appear(s) to be linear from your tables? Explain (using data from your tables) why each is either linear or non-linear.

C. Are you able to write a function (rule, method, procedure) for any of the graphs? If so, write the appropriate function(s) in equation form and explain each part of your equation.

D. Write a description of each graph including whether it is linear or non-linear, areas where the graph is increasing or decreasing, and the comparative rates of increase or decrease.

E. Can you give a physical explanation for why some variables are initially linearly related, but then become non-linear?

Part 2.

A. Notice that **Graph 1** shows power as a function of time and that **Graph 2** shows oxygen consumption rate as a function of power. It is possible from these graphs to develop a graph of oxygen consumption rate as a function of time for this particular treadmill test. Describe why this is possible and how you might go about developing the graph.

B. Similarly, using **Graph 3** you can develop a graph of the rate of lactic acid released as a function of time. Describe how you would develop this graph.

C. Would the oxygen consumption rate as a function of time graph be linear or non-linear? Would it be increasing or decreasing? How can you tell? Describe the graph of the rate of lactic acid released as a function of time in terms of linearity and slope.

D. Produce the graph of oxygen consumption rate as a function of time.

Exercise Target Heart Rate

Brian Sharkey in *Physiology of Fitness* states that:

> The benefits of aerobic exercise and fitness include
> - improved circulation, respiration, and fat metabolism;
> - reduced stress levels, body fat, and risk of heart disease;
> - stronger bones, ligaments, and tendons;
> - weight control;
> - more energy and less fatigue;
> - enhanced mood, self-concept, and body image;
> - greater emotional stability, and
> - a more positive outlook. (1990, p. 2)

He also cites recent polls showing an increase in the number of adults engaging in aerobic activities and a decline in the incidence of heart disease. In order to become aerobically fit in a health-related way, the American College of Sports Medicine (ACSM) recommends exercising 3 to 5 times per week, for 15 to 60 minutes, at your determined Target Heart Rate (THR).

A Target Heart Rate (THR) which provides these benefits is determined using the following. Note that heart rates are measured in beats per minute.

First: You need to determine your resting and maximum heart rates:

resting heart rate: normal heart rate at rest (for example when you first awake in the morning)

maximum heart rate: 220 minus your age (for a person of average fitness)

Second: Now you can determine your THR. Your THR should be calculated using an exercise intensity of between 65 and 90 percent using the following formula:

THR = [(max HR - resting HR) x (percent exercise intensity)] + resting HR

Part 3.

A. To calculate your resting heart rate, you measure your heart rate for three mornings before rising out of bed and take an average of your measured heart rates. If your measured heart rates were 60, 58, 61 beats per minute, what is your resting heart rate?

B. If you are 19 years old and of average fitness, what is your maximum heart rate? Using your own age, what is your maximum heart rate for exercise? How do you think your fitness level would affect your maximum heart rate?

C. For this requirement use your own maximum heart rate and the resting heart rate from part A. Suppose you wanted a moderately hard workout at 70% intensity. What would be your target heart rate? And if you wanted a very hard workout at 85% intensity, what would be your target heart rate?

D. Make a graph of your THR versus your desired exercise intensity percentage using your maximum heart rate and the resting heart rate from part (A).

E. Explain what this graph is showing. Describe the graph in terms of increasing or decreasing, linear or non-linear, and steepness or flatness.

F. As you get older will your THR change? Assuming your resting heart rate stays the same and you plan to exercise at 70% intensity, graph your THR for the next 10 years.

Measuring Aerobic Fitness

Aerobic fitness can be determined in a laboratory test (using a bicycle, treadmill, etc.) which measures maximal oxygen consumption (VO_2 max). Common scores of oxygen consumption range from 3 to 4 liters of oxygen per minute (L/min), with endurance athletes reporting 5 to 6 L/min. Because this measure indicates only total capacity of the cardiovascular system, larger individuals tend to have higher scores. In order to eliminate the influence of body size, "aerobic power" is determined by dividing one's VO_2 max by one's weight in kilograms and translating to ml/(kg x min.).

The following table provides a fitness comparison of college students (ages 18 - 22) from those who are untrained to those who are world-class athletes.

Table 1 - Fitness Comparison

SUBJECTS (fitness level)	MEN [ml/(kg x min)]	WOMEN [ml/(kg x min)]
Untrained	45	39
Active	50	43
Trained	57	53
Elite	70	63
World Class	80+	70+

Part 4.

A. Assume a 20-year old weighing 60 kg has a VO_2 max of 3.3 L/min. What is this person's aerobic power? Describe what this value means. Interpret this person's probable level of fitness if he is a man; if she is a woman.

B. From the data provided, determine a scale for the level of fitness (untrained to world class) and represent the data from Table 1 in graphical form. Explain your choice of the independent and dependent variables.

Part 5.

A. Given the following graphs, match each of the two written situations/descriptions to the appropriate graph. in each graph the horizontal axis represents time and the vertical axis represents rate of oxygen intake.

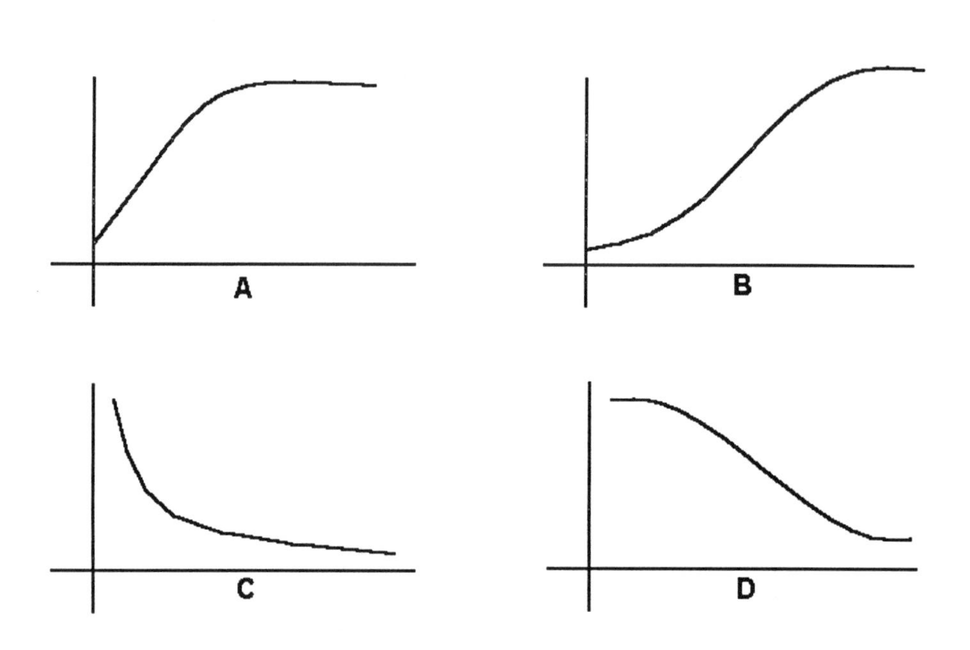

______ As exercise begins, oxygen intake does not meet demands. Oxygen intake must quickly increase, and, when it meets the demand, a steady state is achieved.

______ After exercise, recovery is initially quick and then slows as oxygen returns to resting levels.

B. Using your choices above, complete the graph below, showing oxygen intake in liters per minute as a function of time in minutes for a full exercise session from beginning to recovery. Completely label the graph from the situation information.

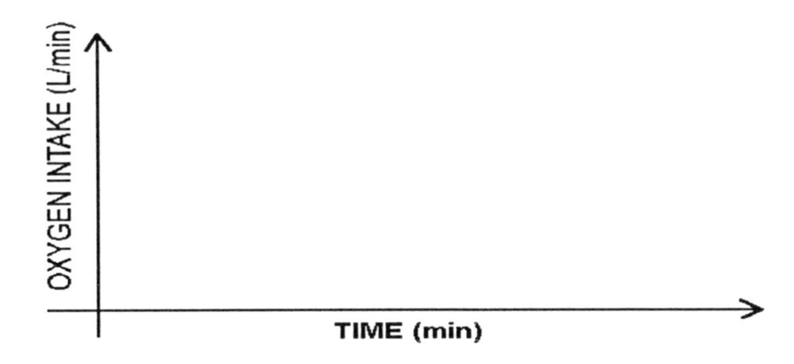

Exercise, Energy, and Fat

Young men average 12.5% body fat, while the average for young women is 25%. Consider that fat is our most abundant source of energy, and most of us have more than we need. What we need to do is to learn how to burn fat during exercise. By doing so we can improve our fitness level and improve our overall health.

As exercise intensifies, the body requires more oxygen. Fatty acid is oxygen poor while carbohydrates carry more oxygen. Therefore, as more oxygen is required, the body switches from fat as a predominant source of energy to a fat-carbohydrate (glycogen) mixture. As length of duration of exercise increases, fat utilization increases with time. During the first half-hour of exercise fat mobilization is delayed, but, as duration of activity increases, fat usage increases.

Part 6.

Given the following graphs, match each of the four written situations/descriptions to the appropriate graph. Variables are shown in parenthesis (input, output).

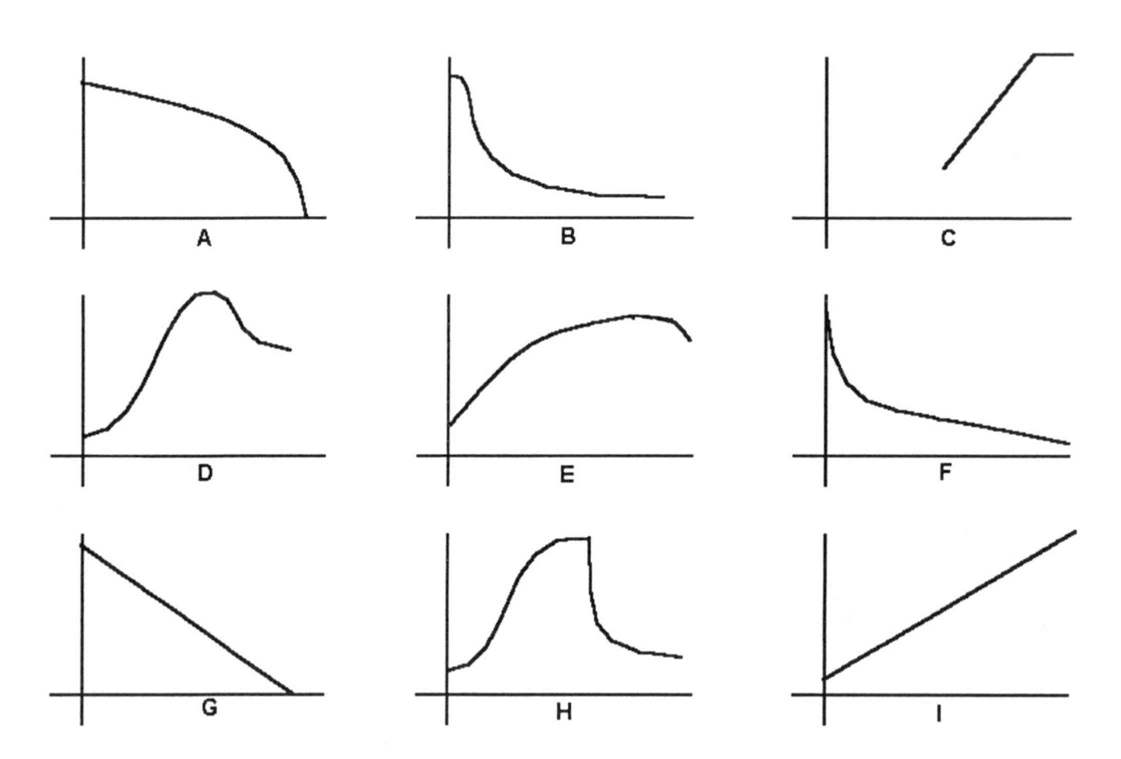

_____ There is a linear inverse relationship between use of fats and carbohydrates in exercise. As exercise increases in intensity, more carbohydrates and less fats are used. (fats, carbohydrates)

_____ 100% of the capacity of our energy system is supported by ATP and its backup, high energy creatine phosphate (CP), at the start of exercise, but after a short time (within 30 seconds) this support quickly drops and other energy systems take over. (time, % of energy system supported by ATP & CP)

_____ The fat and glycogen system steadily increases from a low percentage of the system used to 100 % used. After a long period of exercise this system is not able to continue supporting activity and

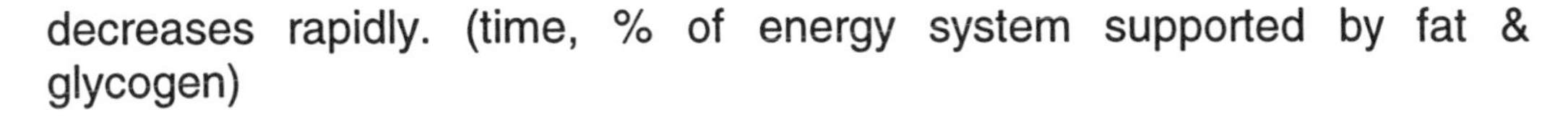

decreases rapidly. (time, % of energy system supported by fat & glycogen)

_____ The glycolysis system starts slowly and then rapidly increases to 100% utilization within 2 minutes of starting exercise. Its support ability is expended quickly, so, once it reaches 100%, it slowly decreases its percentage of capacity to provide energy. (time, % of energy system supported by glycolysis)

_____ The glucose system kicks in after a long duration of exercise and steadily increases until 100% of its capacity is utilized. (time, % of energy system supported by glucose)

Situation 2: The Physical Education Department now wants to interpret the data they obtained from student fitness testing. They have asked you to assist in the analysis.

Part 1.

Before beginning any additional analysis, you need to become more familiar with **Graphs 1 - 3**. You realize that, although the graphs are helpful, you would like to know the relationship between oxygen consumption rate and duration of exercise (time).

A. Conjecture functions for both **Graph 1** and **Graph 2** above. As instructed by your professor, discuss your function conjectures with other members of the class and attempt to reach a class consensus on appropriate functions.

B. After your class has agreed upon the functions, compose (make a composite function of) the two functions. What is the relationship shown by this composite function? Plot the composite function and label the axes appropriately.

C. Compute the total oxygen consumed during this 15-minute treadmill test.

Part 2.

In **Part 1**, we approached the problem qualitatively by trying to fit analytical expressions to curves. We can be more accurate by examining the numerical data provided by the Physical Education Department that was used to generate **Graphs 1** and **2**. With this data, we can more accurately compute the total oxygen consumed.

A. Given the following data (Tables 1 & 2), produce a table of composite data for oxygen consumption rate versus time (Table 3).

Data From Student Treadmill Test

Table 1

Time (min)	Power (watts)
1	14.25
2	28.50
3	42.75
4	57.00
5	71.25
6	85.50
7	99.75
8	114.00
9	128.25
10	142.50
11	156.75
12	171.00
13	185.25
14	199.50
15	213.75

Table 2

Power (watts)	Oxygen Consumption (liters/min.)
10	0.48
20	0.94
30	1.37
40	1.78
50	2.16
60	2.53
70	2.86
80	3.18
90	3.46
100	3.73
110	3.97
120	4.19
130	4.38
140	4.55
150	4.69
160	4.82
170	4.91
180	4.99
190	5.03
200	5.06
210	5.06
220	5.03
230	4.99
240	4.91
250	4.82

Table 3

Time (min)	Oxygen Consumption (liters/min.)

B. Use this new table to find total oxygen consumed. (HINT: Use a suitable numerical integration technique.)

Part 3.

Two friends, Megan and Justin, have been working out together to get ready for the 2-mile race. They have similar levels of conditioning, but have different race strategies. Megan likes to start off slowly, increase power during the race, and conclude with peak effort at the end of the race. Justin, on the other hand, prefers to start off fast with peak energy, decrease power during the race, and conclude at a speed determined by his level of fatigue. The two strategies are graphed below.

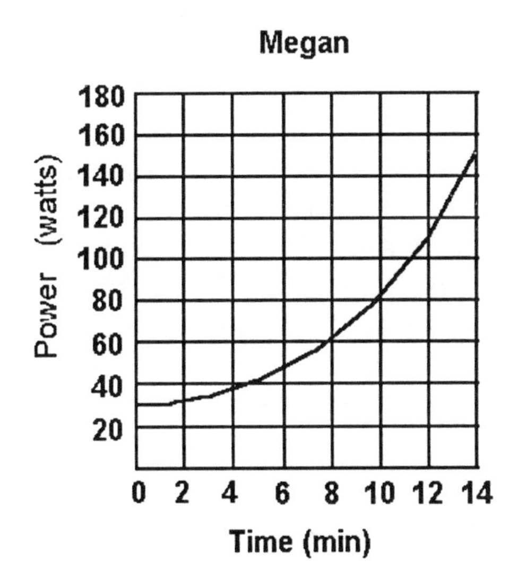

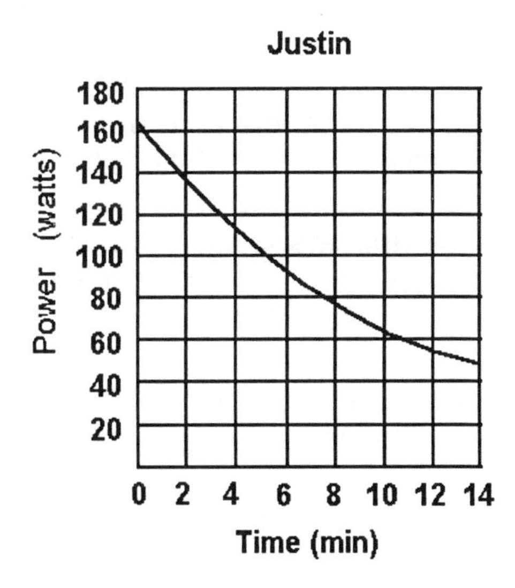

In the following questions, assume that the oxygen consumed versus power data from **Part 2** is valid for this requirement.

A. Determine the total oxygen consumed for each of the two strategies. Which strategy consumes more oxygen? (Hint: You may want to generate several power versus time tables by using the graphs above.)

B. Can you predict which of your friends will have the best race time?

Part 4.

In light of **Parts 1 - 3**, write a short essay describing how you might use Graphs 1-3 (power, oxygen, lactic acid), to find the total lactic acid which accumulated during the 2-mile race. Which race strategy would result in the greater accumulation of lactic acid? Would knowing the amount of lactic acid accumulated change your answer to **Part 3(B)** above? Why?

Title: Getting Fit with Mathematics

SAMPLE SOLUTION

Situation 1. Power and Oxygen Consumption

Part 1.

A. Graphs that are linear are straight lines. Therefore, Graph 1 appears to be linear while Graph 2 does not. Graph 3 looks like it has three linear segments. A graph is linear if for any chosen segment the ratio of the vertical change over the horizontal change remains constant. In Graph 1 it appears that the ratio of the vertical change over the horizontal change is always approximately 14/1. The ratio changes in Graph 2; initially it is 1/20, but then decreases to 1/40. Graph 3 has three segments, each of which appears linear when examined individually. The ratio of the first segment (oxygen consumption rate going from 0 to 2 liters/min) is approximately .5/2 or 1/4. In the second segment (oxygen consumption rate going from 2 to 3 liters/min), the ratio appears to be 1.5/1 or 3/2. Finally, the ratio in the third segment (oxygen consumption rate going from 3 to 4 liters/min), is approximately 8/1.

B. Tables of input and output values for Graphs 1 - 3.

GRAPH 1

INPUT	OUTPUT	ΔI	ΔO	ΔO/ΔI
1	14			
2	28	1	14	14
3	42	1	14	14
4	57	1	15	15
5	70	1	13	13
6	84	1	14	14
7	98	1	14	14
8	112	1	14	14
9	127	1	15	15
10	141	1	14	14
11	155	1	14	14
12	170	1	15	15
13	184	1	14	14
14	199	1	15	15
15	213	1	14	14

GRAPH 2

INPUT	OUTPUT	ΔI	ΔO	ΔO/ΔI
0	0			
20	0.90	20	0.90	0.0450
40	1.75	20	0.85	0.0425
60	2.50	20	0.75	0.0375
80	3.20	20	0.70	0.0350
100	3.70	20	0.50	0.0250
120	4.20	20	0.50	0.0250
140	4.50	20	0.30	0.0150
160	4.75	20	0.25	0.0125
180	4.95	20	0.20	0.0100
200	5.05	20	0.10	0.0050

GRAPH 3

INPUT	OUTPUT	ΔI	ΔO	ΔO/ΔI
0	1.000	0.5		
0.5	1.125	0.5	0.125	0.25
1	1.250	0.5	0.125	0.25
1.5	1.375	0.5	0.125	0.25
2	1.500	0.5	0.125	0.25
2.5	2.250	0.5	0.750	1.50
3	3.000	0.5	0.750	1.50
3.5	7.000	0.5	4.000	8.00
4	11.000	0.5	4.000	8.00

The table of input and output valuess support the conclusions made in **Part 1(A)** above. The table for Graph 1 fluctuates from a ratio of 13 to a ratio of 15, but the most common ratio is 14. It is difficult to read the exact number values from the graph, so it's possible all of the ratios should be 14. If we assume that all the ratios are 14, then the graph is linear.

The table for Graph 2 shows that the ratio of vertical change to horizontal change is decreasing. The graph also shows this as it flattens out toward 200. This graph is not linear.

Graph 3 appeared to have three linear segments. The table also shows three distinct segment with three different ratios of vertical change to horizontal change.

C. We can write linear rules for Graphs 1 and 3. Eventually, we may be able to write a non-linear rule for Graph 2, but we will not do so now.

$$\text{Graph 1: } P = 14(T)$$

This equation says that the Power, P, is equal to 14 times the time, T. It also shows that the slope of the line is 14, the ratio of vertical change to horizontal change.

$$\text{Graph 3:} \quad L = \begin{cases} 1.0 + .025(O) & 0 \le O < 2 \\ -1.5 + 1.5(O) & 2 \le O < 3 \\ -5.0 + 8.0(O) & 3 \le O < 4 \end{cases}$$

This is a "piece-wise" defined graph. The equation says that the lactic acid release rate, L, is dependent upon the oxygen consumption rate, O. Depending upon the oxygen consumption rate, the lactic acid release rate is determined by the function above. The values 0.25, 1.5, and 8.0 are the respective slopes of each segment.

D. The graphs can be described as follows.

Graph 1: This is an increasing linear graph. It is increasing at a constant rate of 14 watts of power for each minute of time.

Graph 2: This is an increasing non-linear graph. The rate of increase is slowing down. This means that the ratio of vertical change to horizontal change is getting smaller (decreasing). The slope of the graph is initially steep and then flattens out.

Graph 3: This is a piece-wise defined graph with three linear pieces. Each piece is increasing. The first segment increases at a constant rate of 0.25; the second at a constant rate of 1.5; and the third at a constant rate of 8.0. These rates tell us the millimoles of lactic acid released per liter of oxygen consumed.

E. In Graph 1 we see that there is a linear relationship between time and power. Recall that the scenario in this case was a treadmill test. The treadmill is gradually inclined and its speed is gradually increased as time goes on. In order to "keep up with" the treadmill, one must increase one's power exertion proportional to the increased inclination and speed. This proportionality causes the linearity in this situation.

Graph 2 appears to be linear at first, indicating there might be a proportional relationship between power and the oxygen consumption rate during the treadmill test. Quickly, however, we can see it is not a constant proportionality and therefore not linear. Physically, it appears the student requires successively smaller oxygen consumption rate increases for constant increases in power (The function is still increasing, but not as quickly.). We could interpret the graph as initially showing one trying to "catch" one's breath going from a stationary state to running on the treadmill. During this period the rate of oxygen consumption increases rapidly. As exercise continues, one does "catch" one's breath and more efficiently increases the oxygen consumption rate relative to the amount of power required.

As a person is exercising, Graph 3 indicates that there are specific rates of oxygen consumption which cause a change in the relationship between the oxygen consumption rate and the lactic acid release rate. One of these points was discussed in the background information as the Onset of Blood Lactic Acid (OBLA).

Part 2.

A. We can develop a graph of the oxygen consumption rate as a function of time. Since Graph 1 shows us a relationship between power and time, and Graph 2 shows us a relationship between the oxygen consumption rate and power, we can use the common variable of power to graph a relationship between time and the oxygen consumption rate.

First, set up a coordinate system with the oxygen consumption rate along the vertical axis and time along the horizontal axis. Select a point from Graph 1-- for example, time = 2, power = 28. Now look at Graph 2 to find the oxygen consumption rate when power = 28. It is approximately 1.2. You are now able to plot your first point on your new graph. The point is at time = 2 minutes, and the oxygen consumption rate = 1.2 liters/min. A similar procedure will yield additional points for your new graph.

B. As above, we can develop a graph of the lactic acid release rate as a function of time using Graph 3 and the new graph from Part 2(A). First we would again set up a coordinate system with the lactic acid release rate along the vertical axis and time along the horizontal axis. From the graph of oxygen consumption rate as a function of time, we would select a point (t, o). Using that point we would go to Graph 3 and determine the lactic acid release rate, l, for the respective oxygen consumption rate, o. Using these two values, we can plot the point (t, l) [indicating time and lactic acid release rate] on the graph of the lactic acid release rate as a function of time.

C. Neither of the two graphs produced in the requirements above would be linear. Both, however, will be monotonically increasing. To determine this, we need to examine where the graphs are coming from. Perhaps putting the information in tabular form would assist in our analysis. As an example, we can look at the data available for the first graph and use that information to produce a graph for the next requirement.

D. We can produce the graph of oxygen consumption rate as a function of time as follows:

FIRST DETERMINE POINTS--->THEN ANALYZE SLOPE

Time	Power	Oxygen
1	14	0.6
2	28	1.2
3	42	1.8
4	57	2.4
5	70	2.8
6	84	3.2
7	98	3.6
8	112	4
9	127	4.3
10	141	4.5
11	155	4.7
12	170	4.85
13	184	4.95
14	199	5

Time	Oxygen	ΔI	ΔO	$\Delta O/\Delta I$
1	0.6	1		
2	1.2	1	0.6	0.6
3	1.8	1	0.6	0.6
4	2.4	1	0.6	0.6
5	2.8	1	0.4	0.4
6	3.2	1	0.4	0.4
7	3.6	1	0.4	0.4
8	4	1	0.4	0.4
9	4.3	1	0.3	0.3
10	4.5	1	0.2	0.2
11	4.7	1	0.2	0.2
12	4.85	1	0.15	0.15
13	4.95	1	0.1	0.1
14	5	1	0.05	0.05

Therefore, oxygen consumption rate as a function of time produces a non-linear increasing graph. The slope of the graph, however, is decreasing. This means it would get flatter as time increased. We can use the new points we've determined to sketch the graph:

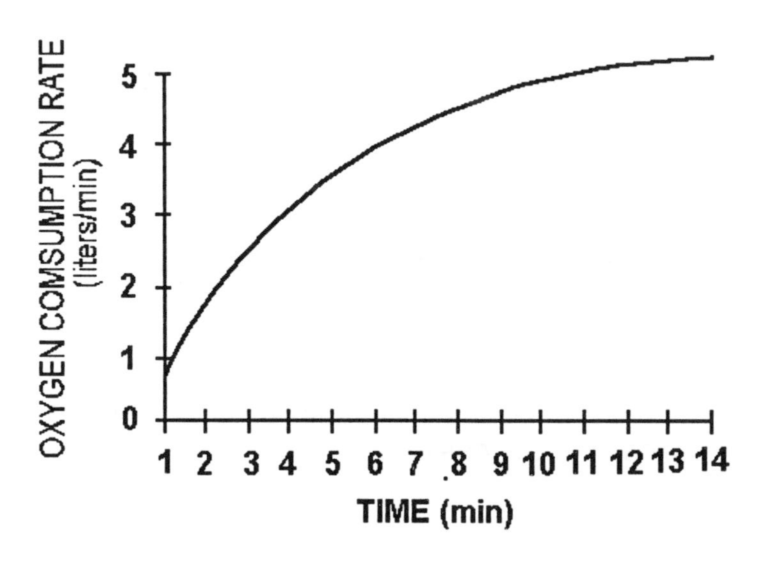

Part 3.

A. The resting heart rate is the average of the three measured heart rates; 60, 58, and 61. Therefore, this person's resting heart rate would be:

$$(1/3)(60 + 58 + 61) = 59.67 \text{ or approximately } 60 \text{ beats/minute.}$$

B. A 19-year old of average fitness would calculate his/her maximum heart rate as:

$$220 - 19 = 201 \text{ beats/minute.}$$

For the student's own age, just substitute the student's age into the formula [220 - age] to obtain the student's maximum heart rate. Recall these are formulas for a person of average fitness. For exercise purposes, a less fit person should use a maximum heart rate less than one calculated by this formula. An extremely fit person may be able to exercise at a higher heart rate and therefore may use a maximum heart rate higher than this formula indicates.

C. We will use a 19-year old with a maximum heart rate of 201 beats per minute to complete the solution. For a moderate workout at 70% intensity we determine that the target heart rate would be:

THR = [(201 - 60) x (.70)] + 60 = 158.7 or 159 beats/minute.

If we wanted to have a very hard workout at 85% intensity, our target heart rate would be:

THR = [(201 - 60) x (.85)] + 60 = 179.85

or approximately 180 beats/minute. This means that to obtain the exercise benefit we desire, our heart rate must be at 159 or 180 beats/minute respectively for 15 to 60 minutes of our exercise period.

D. For the 19-year old in this example solution, the following is a graph of THR as a function of exercise intensity desired.

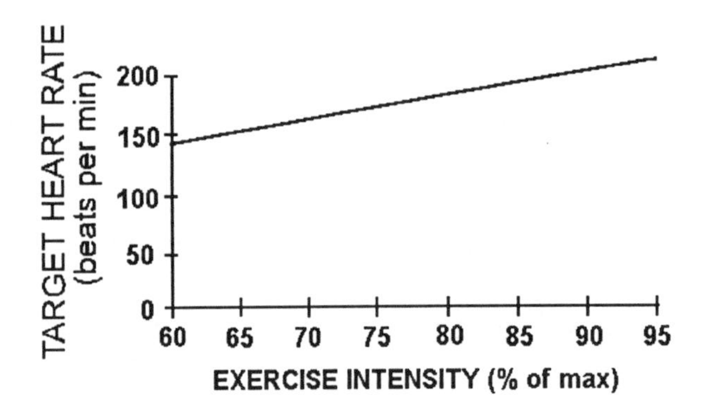

E. The graph shows that there is a linear relationship between THR and exercise intensity. The graph is linear and is increasing. The ratio of change in THR over the change in exercise intensity is constant. This ratio is the rate of change, or slope, of the graph. The slope of this graph is 7 beats per minute/% of max intensity.

F. Assuming our resting heart rate stays the same (60 beats/minute), we can determine our THR at 70% exercise intensity for the next 10 years by using:

$$THR = [(220 - age - 60) \times (.70)] + 60$$

$$THR = [(220 - 60) \times (.70)] + 60 - (age) \times (.7)$$

$$THR = 172 - 0.70\ (age)$$

This equation depicts THR as a function of age. Because the coefficient in front of the independent variable *age* is negative, it indicates this linear function is decreasing. The ratio of change in THR over change in age is a constant (- 0.7) beats/minute. The graph of this function is shown below.

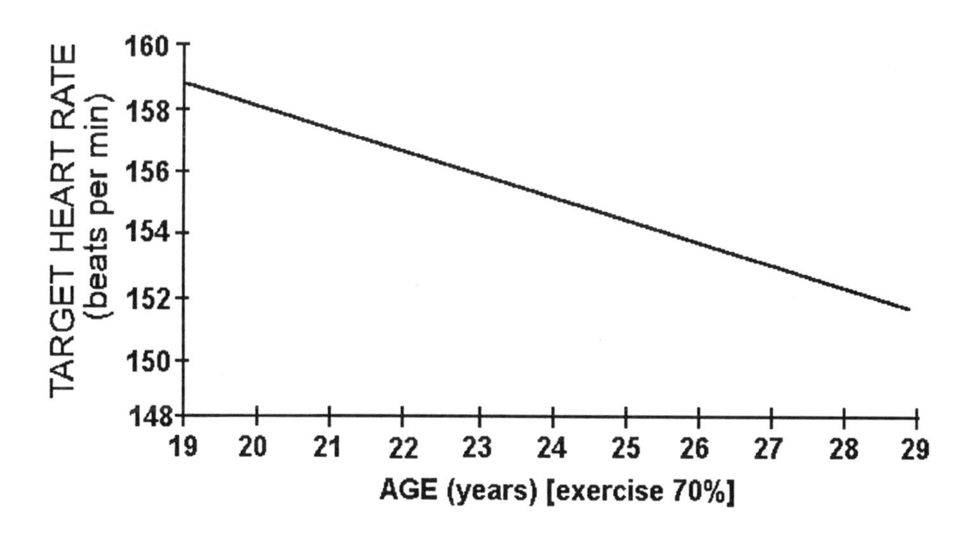

Measuring Aerobic Fitness

Part 4.

A. A person's aerobic power is determined by dividing their VO_2 max by their weight in kilograms. This 20-year old's aerobic power would be determined as:

$$[(3.3 \text{ L/min}) / (60 \text{ kg})]\ [1000 \text{ ml/L}] = 55 \text{ ml/kg min}$$

This value indicates this individual's ability to consume oxygen per unit of body size. The higher the value the more efficient the consumption of oxygen, and the better the level of the individual's fitness condition. If this person is a male, then he is probably an active person, but does not train regularly. If this person is a female, then she probably trains regularly (3 - 5 times per week) for 15 - 60 minutes at her determined THR.

B. We can scale the level of fitness using successively increasing values. For example, we can let 0 = Untrained, 1 = Active, 2 = Trained, 3 = Elite, and 4 = World Class. Or, we can qualitatively assign values to show a greater distinction among the categories. For example, we can let 0 = Untrained, 2 = Active, 4 = Trained, 8 = Elite, and 12 = World Class. By doing this we are showing a difference in degree between the levels. Using the first scale the resulting graph is shown below. We will assume there is a continuum from one level to the next, so we are able to connect the data to form a curve.

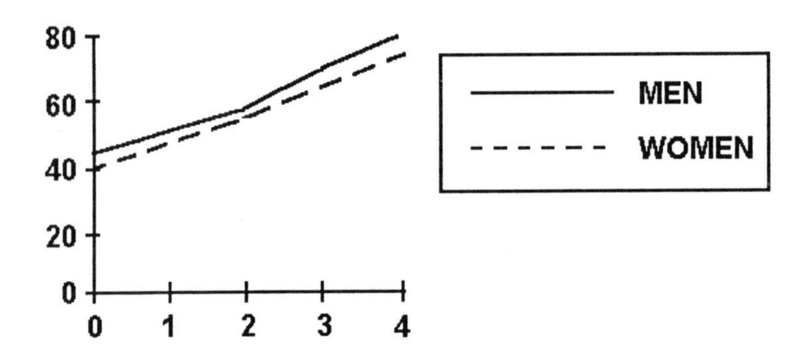

Part 5.

A.

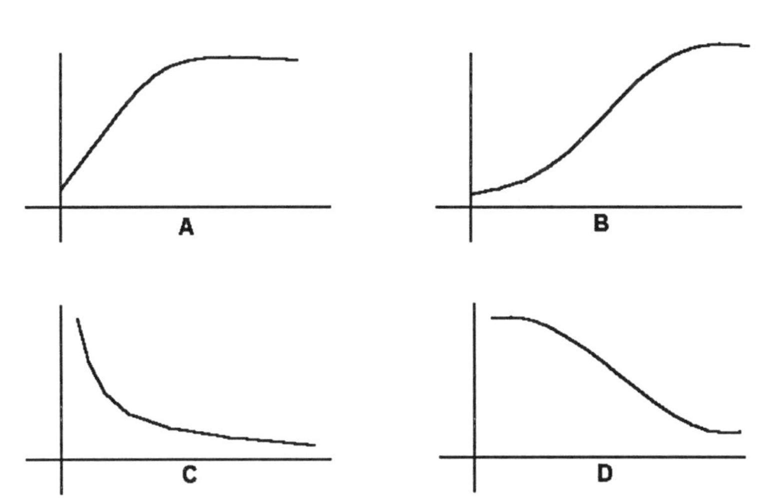

The first situation/description is:

As exercise begins, oxygen intake does not meet demands. Oxygen intake must quickly increase, and, when it meets the demand, a steady state is achieved.

This statement indicates that oxygen intake is the dependent variable and time is the independent variable. The graph will be increasing as oxygen intake goes from a normal/resting amount to the steady state amount needed for exercise. The increase may start off moderate, but quickly becomes rapid, indicating the curve will be steep. As the demand for oxygen is met, the curve will flatten out. Therefore, the slope (the ratio of change in oxygen intake over change in time) will initially increase rapidly, but will then decreases as exercise continues (time increases) and steady state is achieved. Graph A best depicts this situation.

The second situation/description is:

After exercise, recovery is initially quick and then slows as oxygen returns to resting levels.

This statement indicates an initial quick drop and then a gradual, slow, movement of oxygen intake from a high exercise level to a lower resting level. Therefore, we immediately know the graph will be decreasing. Additionally, after an initial quick change (drop), the curve will show a slower decrease, indicating the slope (ratio of change in oxygen intake over change in time) will be increasing as time goes on. [Recall in this situation where the slope is negative because the graph is decreasing, an increasing slope indicates a concave up curve. For example, the slope may change from -5 to -2, indicating it is going from steeper to flatter.] Graph C best depicts this situation.

B. Using the choices above, a graph from Sharkey's *Physiology of Fitness* (page 281) fully depicts the relationship between oxygen intake and time.

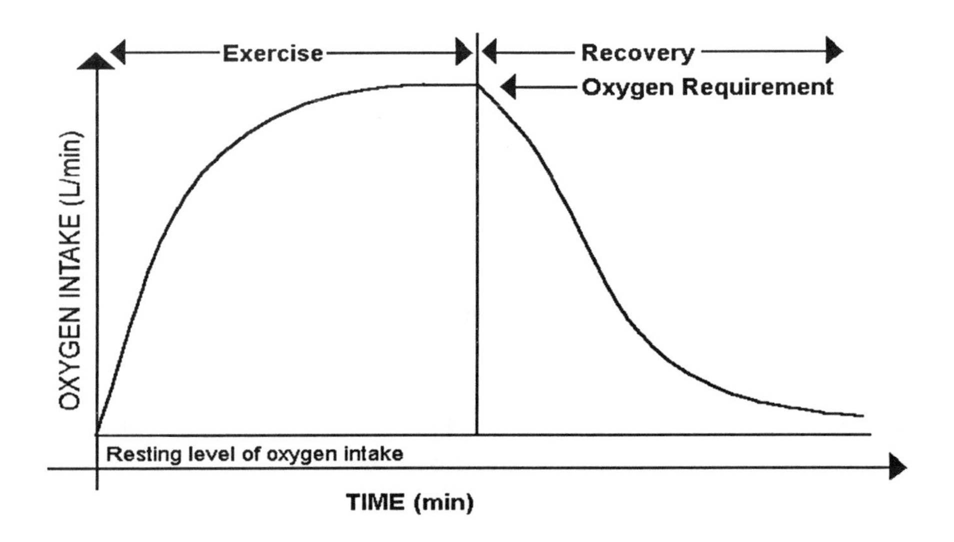

Exercise, Energy, and Fat

Part 6.

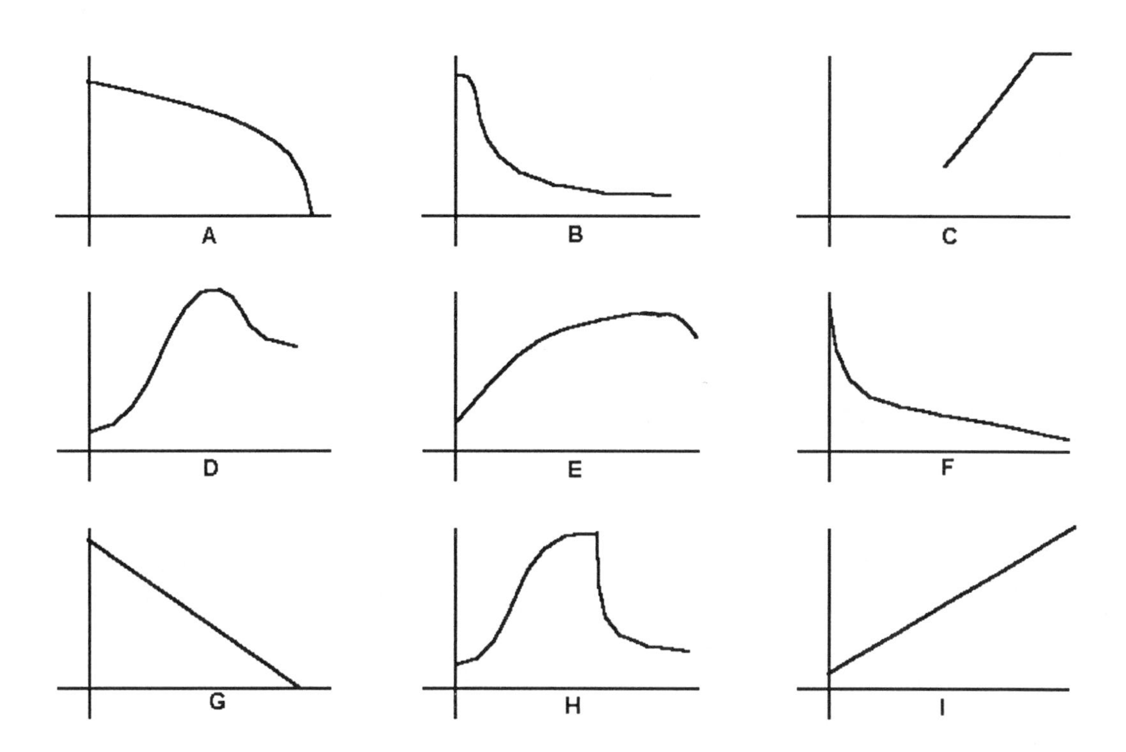

__G__ There is a linear inverse relationship between use of fats and carbohydrates in exercise. As exercise increases in intensity, more carbohydrates and less fats are used. (fats, carbohydrates)

__B__ 100% of the capacity of our energy system is supported by ATP and its backup, high energy creatine phosphate (CP), at the start of exercise, but after a short time (within 30 seconds) this support quickly drops and other energy systems take over. (time, % of energy system supported by ATP & CP) [Choice "F" could also be considered, but it shows an immediate drop of support versus the short delay seen in "B."]

__E__ The fat and glycogen system steadily increases from a low percentage of the system used to 100 % used. After a long period of exercise this system is not able to continue supporting activity and decreases rapidly. (time, % of energy system supported by fat & glycogen) [Both "D" and "H" might also be considered for this description, but both graphs indicate a more fluctuating increase without completely addressing the support of the system over a "long period of time."]

__D__ The glycolysis system starts slowly and then rapidly increases to 100% utilization within 2 minutes of starting exercise. Its support ability is expended quickly, so, once it reaches 100%, it slowly decreases its percentage of capacity to provide energy. (time, % of energy system supported by glycolysis)

__C__ The glucose system kicks in after a long duration of exercise and steadily increases until 100% of its capacity is utilized. (time, % of energy system supported by glucose)

Situation 2: Sample Solution

Part 1.

A. Graph 1 appears to be a linear function and so is of the form $y = mx + b$. Since the y-intercept is 0, the equation is simply of the form $y = mx$. It appears the point (14, 200) is on the line, so the slope is 200/14 = 100/7 = 14.2857 watts/minute. Therefore, we can conjecture the line to be approximately $P(t) = 14.3\, t$, where P is power (in watts) and t is time (in minutes).

Graph 2 appears to be either a quadratic, or possibly a logarithmic, function. Using a curve-fitting technique of estimation, conjecture, a calculator function, or a computer program, we can extract some data points from the graph and try to find an equation using regression analysis. Consider the following table extracted from the graph.

Power (watts)	Oxygen Consumption (liters/min.)
0.001	0.001
20	0.9
40	1.75
60	2.5
80	3.2
100	3.75
120	4.2
140	4.55
160	4.8
180	4.95
200	5.1

Estimation can be done by trying the quadratic $O(P) = c_1 P^2 + c_2 P + c_3$, and conjecturing values for c_1, c_2, and c_3.

Additionally, the TI-82™ calculator provides the following equations through curve fitting. First, we use key **STAT** and choice **EDIT** and **1: Edit** in order to enter our data into a list. Then we use key **STAT** and choice **CALC** to perform curve fitting. The choices we may want to try in this case are:
5: LinReg (ax +b), **6: QuadReg**, **7: CubicReg**, **0: LnReg**, **A: ExpReg**, or **B: PwrReg**.

Each of the respective screen outputs is shown below. (Although beyond the scope of this level, where provided by the calculator, "*r*" indicates the correlation coefficient - the goodness of fit with 1.0 indicating an exact fit):

LinReg	QuadReg	CubicReg
$y = ax + b$	$y = ax^2 + bx + c$	$y = ax^3 + bx^2 + cx + d$
a = .0254318182	a = −1.204837 E-4	a = 1.0926573 E-8
b = .7022727273	b = .0495285548	b = −1.23617 E-4
r = .9666254875	c = −.0206293706	c = .0497785548
		d = −.0237762238

LnReg	ExpReg	PwrReg
y = a + b ln x	y = a (bx)	y = a (xb)
a = 1.936540078	a = .1316331534	a = .1311219143
b = .37739393371	b = 1.024849387	b = .7087873492
r = .7593188494	r = .6538632057	r = .9993828503

Both the quadratic and the power regressions seem to offer a fairly good fit. After graphing both in comparison to the data, the quadratic appears to fit the best. The equation we can use for our purposes is:

$$O(P) = -0.00012\,P^2 + 0.05\,P - 0.02\,,$$

where **O** is the oxygen consumption (in liters/minute) and P is the power (in watts).

B. If we compose the two functions from A:

$P(t) = 14.3\,t$ and $O(P) = -0.00012\,P^2 + 0.05\,P - 0.02$

we obtain:

$$O(P(t)) = -0.00012\,(14.3\,t)^2 + 0.05\,(14.3\,t) - 0.02$$

$$O(t) = -0.0245\,t^2 + 0.72\,t - 0.02$$

This composite function describes oxygen consumption as a function of time.

Plotting the composite function we obtain:

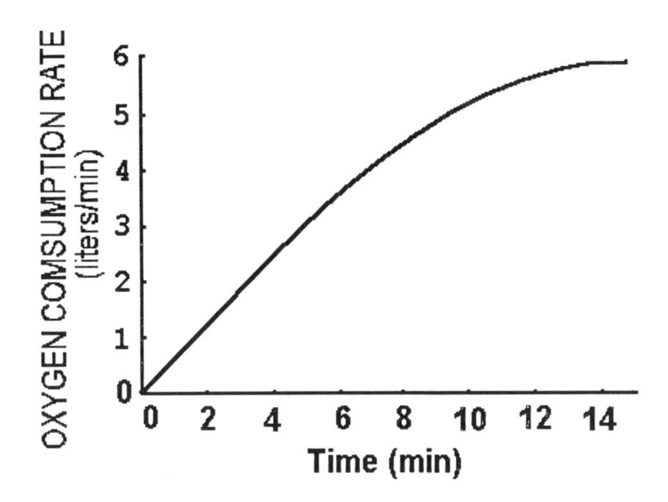

C. To compute the total oxygen consumed, we can integrate our composite function. (We could also approximate the oxygen consumed by simple numerical integration using the graphs we just drew.)

$$\int (-0.0245t^2 + 0.72t - 0.02)dt$$

$$= \left[\frac{-0.0245t^3}{3} + \frac{0.72t^2}{2} - 0.02t \right]$$

= (– 27.5625 + 81 – 0.3) – (0)

= 53.1375 liters.

Part 2.

A. There are at least two ways to complete Table 3.

1. We can use time from Table 1 and with its associated power determine the related oxygen consumed by interpolation of data from Table 2. For example, using Table 1 at t = 1 minute, the power is 14.25 watts. In Table 2 we see that, when power = 10 watts, oxygen consumed is 0.48 liters/minute, and, when power = 20 watts, oxygen consumed is

0.94 liters/minute. So, for time = 1 minute and power = 14.25 watts, we see that oxygen consumed is:

0.48 + [(14.25 -10)/(20 - 10)][0.94 - 0.48] = 0.6755 liters/minute.

Using this procedure we obtain:

Table 3

Time (minutes)	Oxygen Consumption (liters/min.)
0	0
1	0.6755
2	1.3055
3	1.8845
4	2.419
5	2.9
6	3.334
7	3.7233
8	4.058
9	4.3468
10	4.585
11	4.7778
12	4.918
13	5.011
14	5.0585
15	5.0488

2. A second method uses the power and associated oxygen consumption from Table 2, and interpolates to find related time from Table 1 data. For example, in Table 2 when power = 10 watts, oxygen consumed = 0.48 liters/minute. Table 1 shows power = 14.25 watts at time = 1 minute and since it is a linear function, we see $P = 14.25t$, so $t = \frac{P}{14.25}$. Therefore, when power = 10, $t = \frac{10}{14.25} = .7018$ minutes. We could then enter t = .7018 and O = 0.48 into our table.

With this procedure we obtain:

Table 3

Time (minutes)	Oxygen Consumption (liters/min.)
1.7018	0.48
1.4035	0.94
2.1053	1.37
2.802	1.78
3.5088	2.16
4.2105	2.53
4.9123	2.86
5.614	3.18
6.3158	3.46
7.0175	3.73
7.7193	3.97
8.4211	4.19
9.1228	4.38
9.8246	4.55
10.5263	4.69
11.2281	4.82
119,298	4.91
12.6316	4.99
13.3333	5.03
14.0351	5.06
14.7368	5.06
15.4385	5.03
16.1404	4.99
16.8421	4.91
17.5436	4.82

B. To find total oxygen consumed we can use a numerical integration technique. For example, using the data from the Table 3 which we constructed above, we can apply the left- or right-end point rule, mid-point rule, trapezoidal rule, or Simpson's rule. Notice that, if we use the table constructed with time indicated by integers (1, 2, 3, ... minutes), the numerical techniques are easier to apply. Here are some results:

Left-End Point Method:
Total Oxygen= $O(0) + O(1) + O(2) + O(3) + O(4) + ... + O(14) = 48.99$ liters

Right-End Point Method:
Total Oxygen= $O(1) + O(2) + O(3) + O(4) + ... + O(14) + O(15) = 54.04$liters

Mid-Point Method:
Total Oxygen = $O(.5) + O(1.5) + O(2.5) + O(3.5) + ... + O(13.5) + O(14.5)$

But, since we would have to interpolate to get these midpoint values, our estimation from the Mid-Point method will be the same as the Trapezoidal Rule.

Trapezoidal Rule:
Total Oxygen = $(1/2)[O(0) + 2[O(1) + O(2) + O(3) + O(4) + ... + O(14)] + O(15)] = 51.52$ liters

Simpson's Rule:
Total Oxygen = $[\ (15 - 0)/[(6)(15)]\ ]\ [O(0) + O(15) + 2[O(1) + O(2) + ... + O(14)] + 4[O(1.5) + ... + O(14.5)]\ ] = 51.52$ liters

(*NOTE*: You can have students show why the mid-point method and the trapezoidal rules would give the same solution. Additionally, from that comparison they can show why Simpson's rule also gives the same solution in this case.)

Part 3.

A. We have information on Megan's and Justin's power versus time in graphical form for their two-mile race. In order to determine the oxygen consumed, we need to have a table of power versus time for both Megan and Justin. Looking at the graphs provided, we can extract the data and put it into tabular form:

MEGAN	
Time (min)	Power (watts)
0	30
1	31
2	33
3	36
4	40
5	45
6	52
7	60
8	69
9	79
10	90
11	102
12	115
13	130
14	148

JUSTIN	
Time (min)	Power (watts)
0	163
1	152
2	138
3	124
4	110
5	100
6	91
7	83
8	76
9	70
10	65
11	61
12	57
13	54
14	51

Using the table above and the table of power versus oxygen from Requirement 2, we can construct tables of oxygen versus time for Megan and Justin. The procedure is similar to what we did in Part 2.

We find the associated oxygen for the given power by interpolating from Table 2. For example; Megan's power at 1 minute is 31 watts. From Table 2, Part 2, we see that 30 watts indicates oxygen consumption is 1.37 liters/minute and 40 watts indicates 1.78 liters/minute. So, for Megan's 31 watts we can determine her oxygen consumption to be 1.37 + [(31-30)/(40-30)][1.78 - 1.37] = 1.411 liters/minute.

Using the Trapezoidal rule, we can approximate the total oxygen consumed by Megan and Justin as:

Megan: 37.723 liters
Justin: 47.368 liters

MEGAN		
Time (min)	Power (watts)	Oxygen Consumption (liter/min)
0	30	1.37
1	31	1.411
2	33	1.493
3	36	1.616
4	40	1.78
5	45	1.97
6	52	2.234
7	60	2.53
8	69	2.827
9	79	3.148
10	90	3.46
11	102	3.778
12	115	4.08
13	130	4.38
14	148	4.662
15	168	4.892
16	190	5.03

JUSTIN		
Time (min)	Power (watts)	Oxygen Consumption (liters/min)
0	163	4.847
1	152	4.716
2	138	4.516
3	124	4.266
4	110	3.97
5	100	3.73
6	91	3.487
7	83	3.264
8	76	3.052
9	70	2.86
10	65	2.695
11	61	2.563
12	57	2.419
13	54	2.308
14	51	2.197
15	49	2.106
16	47	2.038

To determine who ran the 2-mile race faster we would have to know what the relationship is between Power (watts) and race-speed (miles/hour), or between Oxygen consumption (liters/min.) and race-speed (miles/hour). If, for example, we knew it took a certain amount of oxygen (or power) to run each mile, we could reason that, since Justin consumed more oxygen, he covered more distance in the 14 minutes shown. Since both graphs end at 14 minutes, it is not clear whether or not the two runners completed the 2-mile race within this period.

Consider that the two runners had not finished the race at 14 minutes. Would it be possible for Megan's total oxygen consumption to surpass Justin's if we extend the graphs to perhaps 16 minutes?

The following shows that, even if we extend the graphs, Justin appears to still consume a greater amount of oxygen.

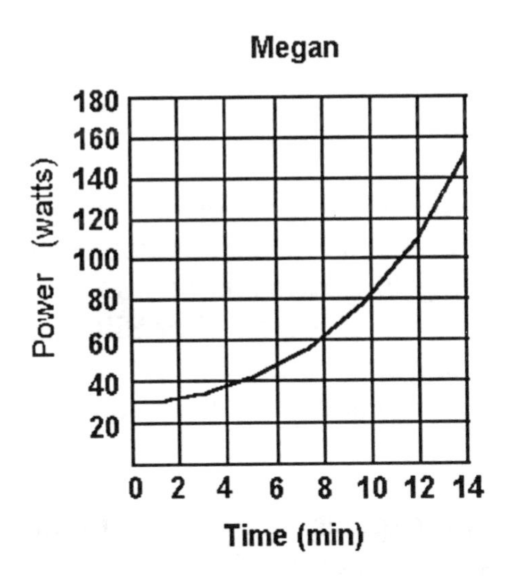

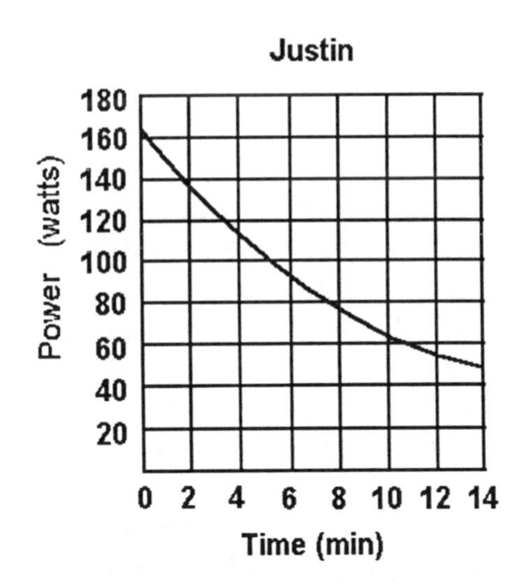

Using the trapezoidal rule we can estimate that after 16 min Megan has consumed 47.4535 liters of oxygen, while Justin has consumed 51.5915 liters. It seems that in order for Megan to consume more oxygen (in this time period), she would have to start out with a greater power level.

Part 4.

Students should write a short essay describing how they could find the total lactic acid accumulated during exercise. Procedures might include:

- Make a table of data from Graph 3 of oxygen consumed (in liter/minute) versus lactic acid released (in millimoles/minute).
- A piece-wise function of lactic acid in terms of oxygen consumption
- Using a function of oxygen consumed in terms of time, form a composite function of lactic acid released in terms of time. Function could be from Requirement 1 as determined by class consensus, or from data on Megan or Justin.
- From the function of lactic acid in terms of time, determine the total lactic acid released using a numerical integration technique.

The results of "cumulative lactic acid" calculation may alter the conclusion in **Part 3(B)** as to the optimal race strategy. A "rabbit" strategy may generate more lactic acid early which must be carried (endured) throughout the race. On the other hand, a "sprint finish" strategy postpones the bulk of lactic acid production until the end of the race.

Title: Getting Fit with Mathematics

Notes for the Instructor

The Getting Fit with Mathematics ILAP is a two-situation project. The first situation is best suited for an algebra or pre-calculus class. It could also be used as a review project at the beginning of a calculus course. The second situation is appropriate for use in a calculus course. For **Situation 2** students should have studied or be currently studying elementary integration and numerical integration techniques. This project's application comes from exercise physiology.

This ILAP is based on aerobic fitness as the capacity of the body to take in, transport, and utilize oxygen. The project is written to be instructive in the discipline of mathematics, as well as informative in the discipline of exercise physiology. The project involves the concepts of oxygen consumption, power, and heart rate during exercise. Solution processes in the first situation include graphing and analyzing graphs, composing functions from graphs, and analyzing linear and nonlinear functions. In the second situation, the techniques of composing functions, elementary curve fitting, elementary integration, and numerical integration are included in the solution processes.

Situation 1.

Parts 1 and 2. The first two requirements involve graphical analysis. Part 1 asks students to determine whether given graphs are linear or nonlinear through the process of making a table of input/output values, and to determine a function for each graph if possible. Part 2 asks students to analyze two graphs which have one variable in common in order to produce a third graph relating the two other variables. Students may have difficulty in forming this relationship and may need either to be walked through the process or to see other examples.

Parts 3 and 4. Parts 3 and 4 ask students to evaluate functions at various values and interpret their solutions. The functions relate to heart rate and aerobic fitness. The functions are provided using a combination of symbols and words. After analyzing the functions, students are asked to produce and analyze graphs related to those functions.

Parts 5 and 6. In Parts 5 and 6, students analyze graphs in order to match graphs to written descriptions of physical situations. In doing so students must determine the independent and dependent variable, and understand the relationship between these variables in terms of linear, nonlinear, increasing, decreasing, slow change, and rapid change.

Although presented as one complete situation in the project, **Situation 1** can be divided and used at separate times during your course. Some Parts may be valuable when studying relationships between functions, while others may be more useful when emphasizing graphing and graphical analysis.

Situation 2.

Part 1. Part 1 is intended to be done as an introduction to the project. In doing Part 1 students become familiar with the situation and the variables involved. Students must qualitatively analyze graphs depicting relationships between exercise duration, power exerted, oxygen consumption rate, and lactic acid release rate. They then conjecture appropriate functions and compose two of their functions. Students are required to integrate their composite function to find total oxygen consumed during a 15-minute treadmill test. You may have students complete this part and compare solutions in class, or do some of Part 1 prior to class and then complete the remaining portion after a class discussion.

Part 2. In Part 2 students are provided with data corresponding to the given graphs. Students use this data to produce a table of composite data for oxygen consumption rate versus time. In order to find the total oxygen consumed during the 15-minute treadmill test, they can use a numerical integration technique.

Part 3 and 4. After looking at the given graphs of the treadmill test both qualitatively and quantitatively, students are provided with two new graphs of two different runners' strategies relating power and time during a race. In Part 3, students must use the results of Parts 1 and 2 to determine the oxygen consumption of the two runners. In Part 4, students must use the results of all three previous Parts to write an essay describing a method to answer additional questions using information which can be obtained from their previous analysis.

Situation 2 should be done as one complete project. In addition to conducting a class discussion during the completion of Part 1, you may want to set up checkpoints during the few days the students are working on the project. Although not required, a graphing package and spreadsheet program are helpful tools for the second situation of this project.

Interdisciplinary Lively Application Project

Title: Decked Out

Authors: Joseph Myers
Cliff Crofford
Kathleen Snook
Terry Mullen

Department of Mathematical Sciences, United States Military Academy, West Point, New York, and Department of Mathematics, Engineering, Physics, and Computer Science, Carroll College, Helena, Montana

Editor: David C. Arney

Mathematics Classifications: Algebra, PreCalculus

Disciplinary Classifications: Engineering

Prerequisite Skills:
1. Solving Systems of Equations
2. Solving Quadratic Equations
3. Solving Linear Inequalities
4. Proportional Rates
5. Elementary Optimization

Physical Concepts Examined: Area, Volume

Materials Available:
1. Problem Statement; Student
2. Sample Solution; Instructor
3. Sample Student Solution (by Brian Niemann); Instructor
4. Notes for the Instructor

Computer Requirements:
1. Graphing Package
2. Spreadsheet (optional)

INTRODUCTION.

The student government at your school wants the administration to approve funding for a deck on the back of the student union building. The outdoor deck would assist in alleviating the crowded conditions in the student union dining facility (at least during nice weather). The deck could be used for meetings, receptions, and other social activities.

Administration officials advised the student government to develop a plan for the project. The only restriction was that no existing trees or shrubs could be removed. A committee was formed to examine the site (see Figure 1) and come up with a general design for the deck. Faculty members with architectural and engineering experience helped the students formulate a preliminary design.

The general design agreed upon (see Figure 2) is actually a system of three separate decks joined together. Because of your mathematics, engineering, and computer experience, you have been asked to help finalize the desk's design, assemble a bill of materials, and prepare a cost estimate for the project. If the cost appears reasonable, the administration will approve the project and have the deck built professionally. The students have agreed to use funds from their activity account to buy the stain for the deck and tables. The students will also apply the stain.

SCOPE OF WORK

Your task is to finalize the design of the deck and prepare a report which will be presented to the administration. The report must be well-written and readable by the audience (administrators at your school). As a minimum, the report should include:

- A professional-looking cover
- A one-page (or less) narrative summarizing the project. (This is called an executive summary).
- All drawings showing the site, deck layout, table arrangements, post holes, decking, and other features you feel are required to describe the project completely.
- A bill of materials showing quantities, costs, and totals.

DESIGN REQUIREMENTS

Prior to determining the bill of materials, you need to refine your design through some additional analysis. Keep in mind both geometrical and algebraic ideas as you perform your analysis and write your report. The figures on the next page provide you with the geometry needed to complete the design and bill of materials for the deck.

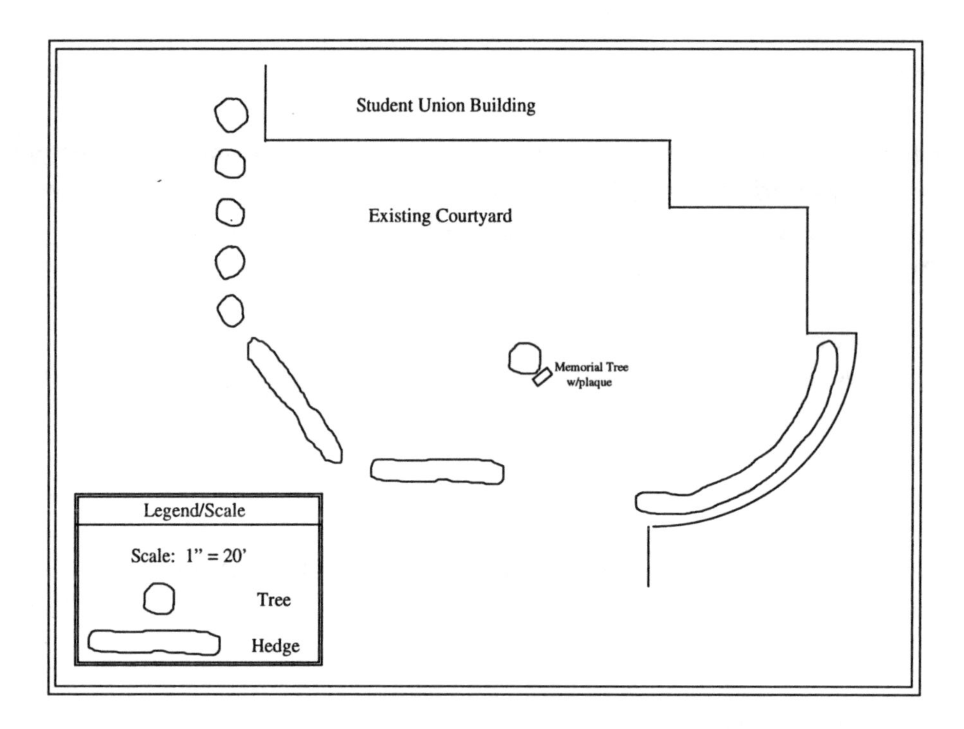

Figure 1: Site Map

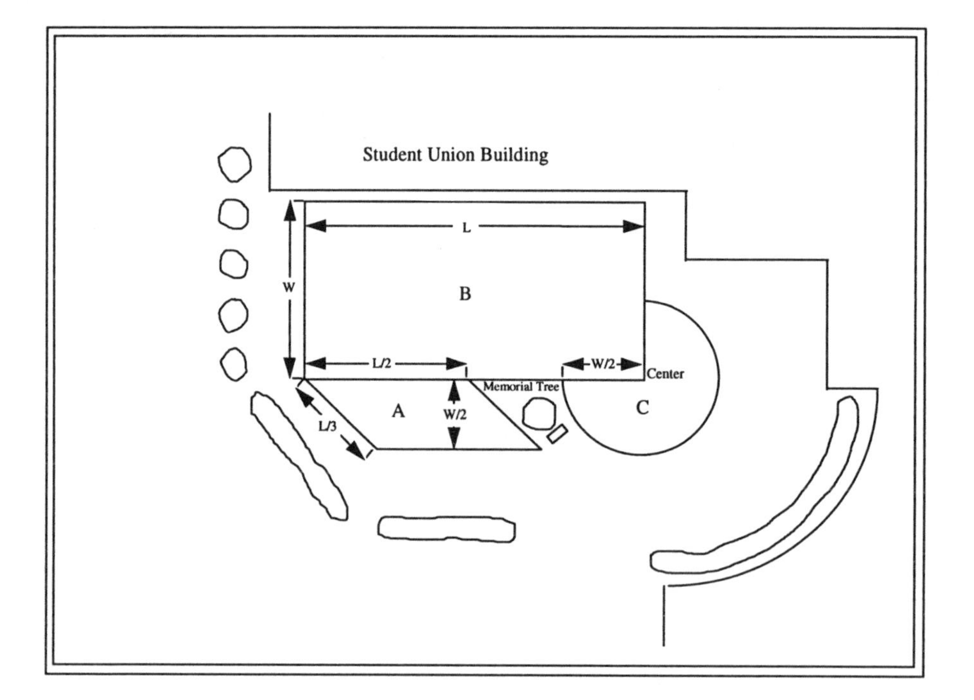

Figure 2. General Design

Part 1. Deck Dimensions and Area

A. The size of decks A and C depend on that of deck B. To satisfy aesthetic proportion requirements, the recommended length of deck B is 1.5 times the width in feet plus 20 feet or $L_B = 1.5(W_B) + 20$. For the projected number of students using the deck, the architect recommended deck B have 2,400 ft.^{2} of area.

1) What are the dimensions (length and width) of deck B? What is its area?

2) Using Figure 2 and the dimensions of deck B, find the dimensions and resulting areas of decks A and C.

B. Will the planned deck with these dimensions fit in the desired location?

C. What is the total area in square feet provided by the deck system?

D. Show the deck (to scale) on your site map.

E. Draw a "close up" of the deck. Include dimensions and label each section's area.

F. The "footprint" of a picnic table is shown in Figure 3. For ease of movement there must be one foot of free space between the benches and any obstruction (for example the wall of the building or the edge of the deck section), and two feet of free space between adjacent picnic tables.

Picnic Table Footprint

Bench (1' x 6')
Table Top (3' x 6')
Bench (1' x 6')

Figure 3. Footprint

How many picnic tables (arranged in a reasonable fashion) can you fit on the deck with no table straddling two separate deck sections? Include a drawing showing the picnic tables placed on the deck system.

According to the purchasing department each picnic table costs $150.00.

Part 2. Concrete

You will need concrete delivered and placed to set the support posts. You decide to check and compare the costs from two different companies. The companies provide you with the following price scales:

Company A		Company B	
Flat Delivery Fee	$35	Flat Delivery Fee	$50
5 cubic yards or less	$10/cy	5 cubic yards or less	$9/cy
over 5 & up to 10 cy	$8/cy	over 5 & up to 8 cy	$7/cy
over 8 & up to 15 cy	$7/cy	over 8 & up to 12 cy	$6/cy
over 15 cy	$6/cy	over 12 cy	$5/cy

A. There are 95 support posts in your deck system, and the local building code requires a minimum post depth of 3 feet. You are going to use 4 x 4- inch posts and drill the post holes with a 10-inch auger (see Figure 4). Determine the amount of concrete (in cubic yards) needed to set the posts at a depth of 3 feet. (NOTE: The actual dimensions of a 4 x 4 post are 3.5 x 3.5 inches.)

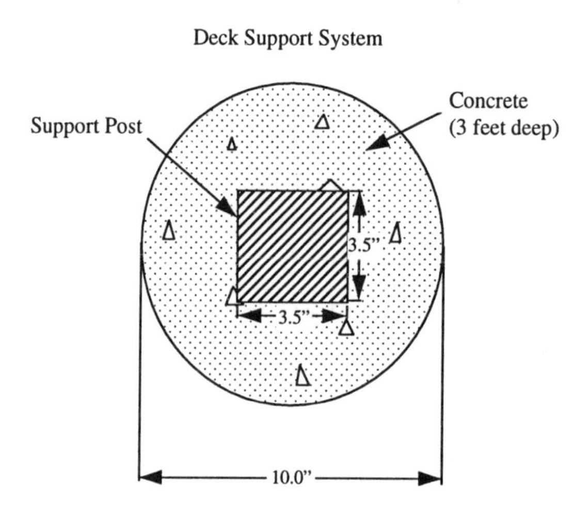

Figure 4. Deck Support

B. Which company will you call to deliver your concrete?

Part 3. The Decking.

A. You are now ready to look at the actual decking. If the 2 x 6-inch decking must be arranged as shown in Figure 5 (with a 1/4-inch space between boards), determine the number and sizes of the boards necessary to deck the entire structure WITH THE LEAST AMOUNT OF WASTE and THE LEAST NUMBER OF BOARDS. A piece of "waste" must be at least 4 feet long to be used in another row of decking. Boards may be placed end to end to span a necessary length. (Assume that the boards for the joists and other support structures are available for free from your school's maintenance facility; do not include them in your analysis.)

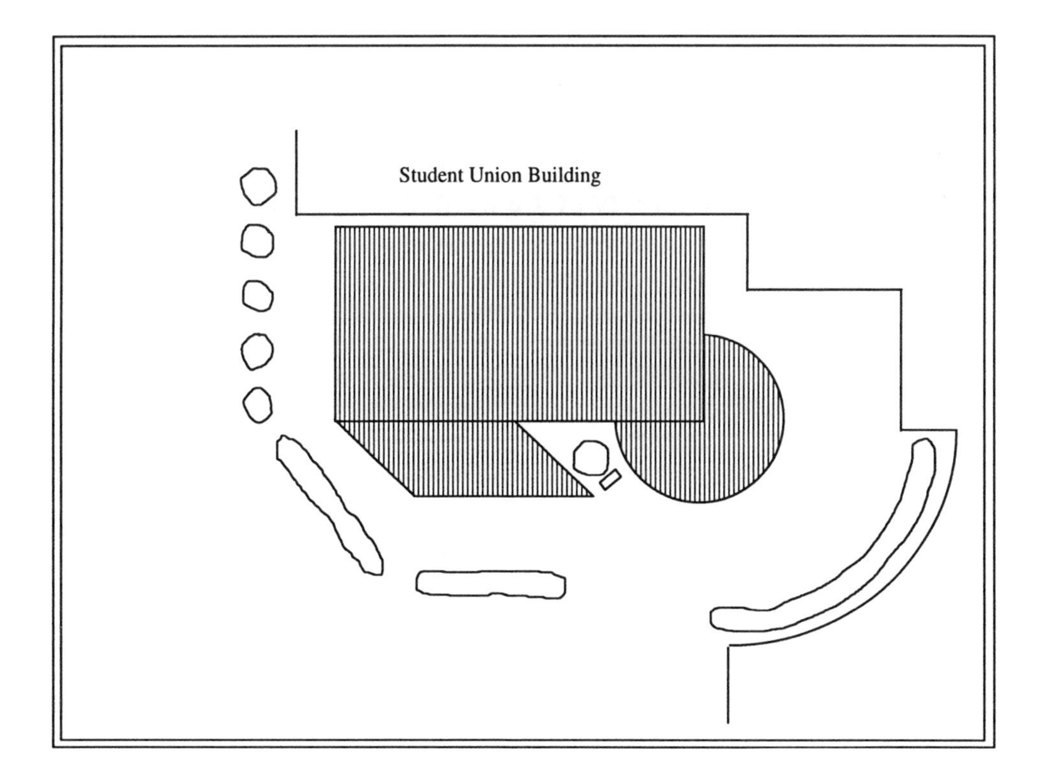

Figure 5. Decking

B. Now find the cost of the decking using the price list below.

Pressure Treated Pine Decking

Board Size (in x in x ft):	**2 x 6 x 8**	**2 x 6 x 10**	**2 x 6 x 12**	**2 x 6 x 16**
Cost:	$5.00	$8.00	$11.00	$16.00

Nominal Size: 1.5 x 5.5 x actual foot length
("nominal" means actual)

BILL OF MATERIALS AND COSTS

Prepare a list of the materials needed to build the deck and purchase the picnic tables. Include the quantity and cost of each item and show the total cost.

EXECUTIVE SUMMARY

Write a one-page (or less) narrative summarizing the important features of the project. Place this summary at the front of your report.

Title: Decked Out

SAMPLE SOLUTION

Part 1. Determining Area Measurements

A. Proportional requirements indicate $L_B = 1.5(W_B) + 20$ feet, and area requirements imply $L_B W_B = 2{,}400$.

1) Using the requirements above we find the dimensions of section B

$$L_B = 2{,}400 / W_B$$

And $$1.5(W_B) + 20 = L_B$$

Therefore, $$1.5(W_B) + 20 = 2{,}400 / W_B$$

$$1.5(W_B)^2 + 20 W_B = 2{,}400$$

$$1.5(W_B)^2 + 20 W_B - 2{,}400 = 0$$

Solving the quadratic equation, we obtain $W_B = -47.2184$ or 33.8851. Choosing the positive solution:

$$W_B = 33.88 \text{ feet}$$

Therefore, $$L_B = 2{,}400 / W_B = 2{,}400 / 33.88 = 70.83 \text{ feet}$$

Letting $W_B = 34$ feet and $L_B = 71$ feet, we meet both the proportionality and area requirements.

2) Since section C is actually three-fourths of a circular deck of radius $0.5\,(W_B)$, then section C is three-fourths of a circle of radius of $0.5\,(34) = 17$ feet centered at the corner of section B. Section A is a parallelogram with both top and bottom sides equal to $0.5\,(L_B)$, so those lengths are $0.5\,(71) = 35.5$ feet. The height of Section A is the same as the radius of section C which is 17 feet. The other two sides are given as $(1/3)(L_B)$, so those lengths are 23.67 feet or 23 feet and 8 inches.

B. The resulting dimensions indicate that there must be at least 71 feet along the building for section B, and there are actually 75 feet.

Additionally, there must be 51 feet of clearance from the building to the southern edge of section A, and there are 55 feet of clearance. The circular section C fits well within the hedges along the circular wing of the student union.

C. The total area of the deck system is equal to the sum of the areas of the three sections:

Section A: Parallelogram -> Area = (35.5)(17) = 603.5 square feet
Section B: Rectangle -> Area = (71)(34) = 2,414 square feet
Section C: Three-Quarter Circle -> Area = $(3/4)(1/2)(\text{pi})(17)^2 = 340.47$

So the total area is: 603.5 + 2414 + 340.47 = 3,357.97 or approximately 3,358 square feet.

F. 1) We can make a square of area 3,358 square feet by making the length of each side of the square equal to the square root of 3,358 or 57.95 feet. Because the length of both sides of a square are equal, it does not make any difference how the tables are placed on the square (vertically or horizontally), we can simply determine how many lengths and how many widths can go along a side of 57.95 feet with the appropriate spacing.

We start with a side of 57.95 feet and take away 1 foot from each end of the side as required. Our picnic tables can then extend across a length of 55.95 feet with at least a 2-foot spacing between adjacent tables. Letting T_n = the number of 5 x 6-foot picnic tables of orientation n, we could write equations for the number of tables that would fit in the 55.95 feet both vertically and horizontally as follows:

Vertical: $5(T_v) + 2(T_v - 1)$ "less than or equal to" 55.95

Horizontal: $6(T_h) + 2(T_h - 1)$ "less than or equal to" 55.95

So, T_v = 8 tables, and T_h = 7 tables. Therefore, on this square we could orient the tables vertically and place 8 tables along the top row and make 7 rows, or orient the tables horizontally and place 7 tables in the top row and make 8 rows. In any case, the number of tables that fit on the square is (7)(8) = 56. Solution Figure 2 depicts both orientations.

2) It is a little more difficult to fit the tables on our designed deck, but we can look at each section separately since no table may sit on two different sections.

Rectangle: Since the sides of a rectangle are not all of equal length, the orientation of the picnic tables may make a difference in the number we can fit on the deck. Start with the tables oriented vertically (ends along the 71-foot side and benches along the 34-foot side). We can write our equations as follows showing the feet required for each table and the 2-foot spacing required between tables (remember the "-2" on the right hand side accounts for the 1-foot spacing required at each end of the row or column):

Vertical: $5(T_V) + 2(T_V - 1)$ "less than or equal to" $71 - 2$

Horizontal: $6(T_h) + 2(T_h - 1)$ "less than or equal to" $34 - 2$

Therefore, $T_V = 10$, and $T_h = 4$ and we can fit $(10)(4) = 40$ picnic tables oriented vertically.

Now let's see how many we can fit on the rectangular section if we orient them horizontally (ends along the 34-foot side and benches along the 71-foot side). The equations become:

Vertical: $5(T_V) + 2(T_V - 1)$ "less than or equal to" $34 - 2$

Horizontal: $6(T_h) + 2(T_h - 1)$ "less than or equal to" $71 - 2$

Therefore, $T_V = 4$, and $T_h = 8$ and we can fit $(4)(8) = 32$ picnic tables oriented horizontally.

We can see it is more efficient to place the tables vertically (short ends toward the student union building) and place 40 tables on deck section B.

Parallelogram: Placing the picnic tables on the parallelogram section A is more difficult. We estimate by looking at a rectangle. Doing that we obtain:

Oriented Vertically: $5(T_V) + 2(T_V - 1)$ "less than or equal to" $35.5 - 2$,

$6(T_h) + 2(T_h - 1)$ "less than or equal to" $17 - 2$,

giving $T_v = 5$, $T_h = 2$ and the number of tables = (5)(2) = 10.

Oriented Horizontally: $5(T_v) + 2(T_v - 1)$ "less than or equal to" 17 - 2,

$6(T_h) + 2(T_h - 1)$ "less than or equal to" 35.5 - 2,

giving $T_v = 2$, $T_h = 4$, and the number of tables = (2)(4) = 8.

Remember that these numbers are only an estimate. Because of the angle of the side of the parallelogram and the requirement that the picnic tables be 1 foot away from the edge of the deck, the number is certainly less than 10 tables. There are two ways to proceed from this point:

a. Use "to scale" templates of picnic tables and the "to scale" diagram of the deck to place the picnic tables with the appropriate spacings.

b. Using measurement of slope (rise over run), calculate where the tables can be placed along the edges of the deck while still maintaining spacing.

In either case (templates being the easier and most visual), we discover there is dead (not usable) space, especially in the acute-angle corners, where tables cannot be placed. It appears we can place 8 picnic tables on section A.

Circle: It is difficult to calculate the number of picnic tables that can be placed on this three-quarter circle. The best method to use may be templates of the picnic tables. Make sure that all spacing requirements are met. There is some dead space around the edges, but we see that 9 tables can be placed on section C.

Therefore the total number of picnic tables is (40) + 8 + 9 = 57.

3) The number of picnic tables found to fit on a square of area 3,358 square feet is one less than the number of picnic tables found by looking at our designed deck It initially appears that they should be different, but intuitively we may think the regularity of the square should provide "room" for more picnic tables. The difference, however, lies in the dimensions of the square. Even though the square is a regular shape, its dimensions are not optimal for this size of picnic table (5 x 6 feet). Just as some space on our irregular parallelogram and circular deck sections was dead space, there is a significant amount of dead space on the square also. The square provided for 8.2785 tables along one side and 7.2438 tables along the other. Since we can have only whole tables, the square provides space for (7)(8) = 56 tables, while at the same time leaving approximately 222 square feet of dead space.

Given that a picnic table of dimension 5 x 6 feet requires, on average, a rectangle of dimensions 7 x 8 feet when we incorporate spacing, we can determine a bound on the number of picnic tables possible on an area of 3,358 square feet as (3,358)/[(7)(8)] = 59.96 or 59 picnic tables (with 0.96 (7)(8) = 53.76 square feet dead space). [Note: Students could explore finding the dimensions and shape of a deck providing space for 59 picnic tables.]

Part 2. Determining Concrete Requirements

A. In order to compare these two companies we must compare them for each interval. Note that the intervals are different for the two companies. For comparison, the intervals and resultant equations can be broken down as follows:

cy concrete ordered	**Company A**	**Company B**
5 cubic yards or less	35 + 10(C)	50 + 9(C)
over 5 & up to 8 cy	35 + 8(C)	50 + 7(C)
over 8 & up to 10 cy	35 + 8(C)	50 + 6(C)
over 10 & up to 12 cy	35 + 7(C)	50 + 6(C)
over 12 & up to 15 cy	35 + 7(C)	50 + 5(C)
over 15 cy	35 + 6(C)	50 + 5(C)

Each interval can now be compared using graphical techniques on paper, calculator, or computer; or by algebraic analysis.

By algebraic analysis, one can check the two end points of each interval. If one company's price is more expensive at both end points, then, since the price equations are linear, that company's price will be more expensive over the entire interval.

Interval End Pts.	Company A	Company B	Choice
(0, 5]	(35, 85]	(50, 95]	Company A
(5, 8]	(75, 99]	(85, 106]	Company A
(8, 10]	(99,115]	(98, 110]	Company B
(10, 12]	(105, 119]	(110, 122]	Company A
(12, 15]	(119, 140]	(110, 125]	Company B
(15, 25]	(125, 185]	(125, 175]	Company B

We determine the surface area of the concrete to be the area of a 10-inch diameter circle minus the area of the 4 x 4 post (nominally 3.5 x 3.5). The surface area is then:

$$\pi\ [0.5(10)]^2 - (3.5)(3.5) = 66.2898 \text{ square inches.}$$

Since each post hole is 3 feet deep, then the concrete required for each post hole will be:

[(66.2898sq in)/(144 sq in / sq ft)][3 ft/hole] = 1.381 cubic feet / hole.

For the 95 support posts, the total amount of concrete will be:

(95 holes) (1.381 cubic feet / hole) = 131.1985 cubic feet.

Since concrete is sold in cubic yards, we must translate to:

(131.1985 cubic feet) / (27 cubic feet / cubic yard) = 4.8592 cubic yards.

So, we will need almost 5 cubic yards of concrete to set the support posts.

B. For 5 cubic yards of concrete, we should use Company A.

Part 3. The Decking

A. Along the 71-foot side, we can determine the number of deck rows (incorporating the 1/4 inch spacing) by the equation:

$$5.5\ (R) + 0.25\ (R - 1)\ \text{"greater than or equal to "}\ (71 \times 12)\ \text{inches}$$

Therefore, R = 163 rows of decking boards. Notice that this number of rows extends 0.33 inches beyond the 71-foot deck. We can either allow the overhang on one end of the deck or slightly decrease a few of the 1/4 inch spaces. A more economical solution would be to use 162 rows and increase the spacing between decking to make up the (5.5 - 0.33) = 5.17 inches from the lost row. In place of a 0.25 inch spacing, we would make the spacing [1/4 + (5.17/161)] = 0.28 inches.

We can set up a table to determine number of boards of each size and the associated waste. With the given orientation, we will be concerned with the lengths needed to cover the 34 foot side.

Board Size	# to span 34 ft	waste in each row
2 x 6 x 8	5	6 '
2 x 6 x 10	4	6 '
2 x 6 x 12	3	2 '
2 x 6 x 16	3	14 '

Recall that there was no restriction on combining different length boards in each row, but only that the waste lumber must be at least 4 feet long to be able to use.

Since we want to use the least number of boards, as well as produce the least waste, we should look at the 16-foot pieces first. The problem with the 16 foot sections is that, after two boards are used, a 2-foot length is needed to get to the end of the 34-foot deck. Therefore, only one complete 16 foot board can be used along with another size board, part of another size board, or another "part" of a 16-foot board. What is needed is combinations of lengths to make up 34 feet of deck. Here are the combinations with one length of 16 feet and all other lengths of at least 4 feet.

16 ft - 14 ft - 4 ft
16 ft - 12 ft - 6 ft
16 ft - 10 ft - 8 ft
16 ft - 10 ft - 4 ft - 4 ft
16 ft - 8 ft - 6 ft - 4 ft
16 ft - 6 ft - 6 ft - 6 ft
16 ft - 6 ft - 4 ft - 4 ft - 4 ft

The following are some examples of combinations (there are certainly others) obtained by maximizing the splitting of 16-foot boards (to minimize total number of boards):

16 foot ** 12 foot ** (3/8) 16 foot and then
16 foot ** (1/2) 16 foot **(5/8) 16 foot and then
16 foot ** (1/2) 16 foot ** 10 foot --> 7 boards in every 3 rows with 0 waste

16 foot ** (1/2) 16 foot ** 10 foot
16 foot ** (1/2) 16 foot ** 10 foot --> 5 boards in every 2 rows with 0 waste

16 foot ** 12 foot ** (1/2) 12 foot
16 foot ** 12 foot ** (1/2) 12 foot --> 5 boards in every 2 rows with 0 waste

16 foot ** 10 foot ** 8 foot --> 3 boards in every row with 0 waste

Since 162 rows are needed, and 162 is divisible by 3, we can make 162 rows out of 54 sets -- our most efficient combination (7 boards every 3 rows).

The bill of materials using this combination would be:

2 x 6 x 10	1/set x 54 sets	54
2 x 6 x 12	1/set x 54 sets	54
2 x 6 x 16	5/set x 54 sets	270
Total Boards	378	

B. The cost for the bill of materials for the boards using only the above plan would be:

$$54\,(8) + 54\,(11) + 270\,(16) = \$5{,}346.00$$

Title: Decked Out

STUDENT SOLUTION

Solution by Brian Niemann, Carroll College, Helena, Montana, as course submission for this ILAP. Brian's instuctor was ILAP author Terry Mullen.

Executive Summary

Now that we are moving into the spring and summer months, the student body is once again becoming more interested in spending time outside. This interest has brought rise to the recurring idea for better outdoor gathering facilities here at the college. The following report outlines the student government's proposal for a deck to be built behind the student union building (site shown in Figure 1) by funds to be approved by the administration. During good weather, it will reduce crowding in the dining facilities and provide a much desired and currently lacking outdoor gathering place for meetings, receptions, or other social events.

The deck will consist of three parts and have a total area of 3,656 square feet (shown in Figure 2). Its size allows for up to 46 picnic tables (shown in Figures 3 and 6), so could comfortably seat about 276 people. In addition, the design avoids the removal of current trees, hedges, bushes and the Memorial Plaque.

The structure itself will be supported by 95 support posts anchored by concrete (see Figure 4). The decking will be mostly 16 foot boards placed parallel to the rear of the building (adjacent rows will of course be offset so joints are stronger). Some shorter boards will of course be necessary to complete some of the smaller lengths but if cut-offs are used elsewhere, board waste should be minimized (see Figure 5, actual board arrangement not shown).

The entire project will cost about $14.5 thousand (see bill of materials and cost below). This price includes concrete, support posts, decking boards and tables.

Bill of Materials and Costs

Item	Quantity	Unit Price	Cost
Support Posts	95	$ 5.00	$ 475.00
Picnic Tables	46	$ 150.00	$ 6,900.00
Concrete*	4.86cy	$ 10.00/cy	$ 83.60
16-foot boards	292	$ 16.00	$ 4,672.00
12-foot boards	48	$ 11.00	$ 528.00
10-foot boards	123	$ 8.00	$ 984.00
8-foot boards	211	$ 5.00	$ 1,055.00
Total:			**$14,697.60**

*Concrete will be purchased from Company A at $10/cy with a delivery charge of $35

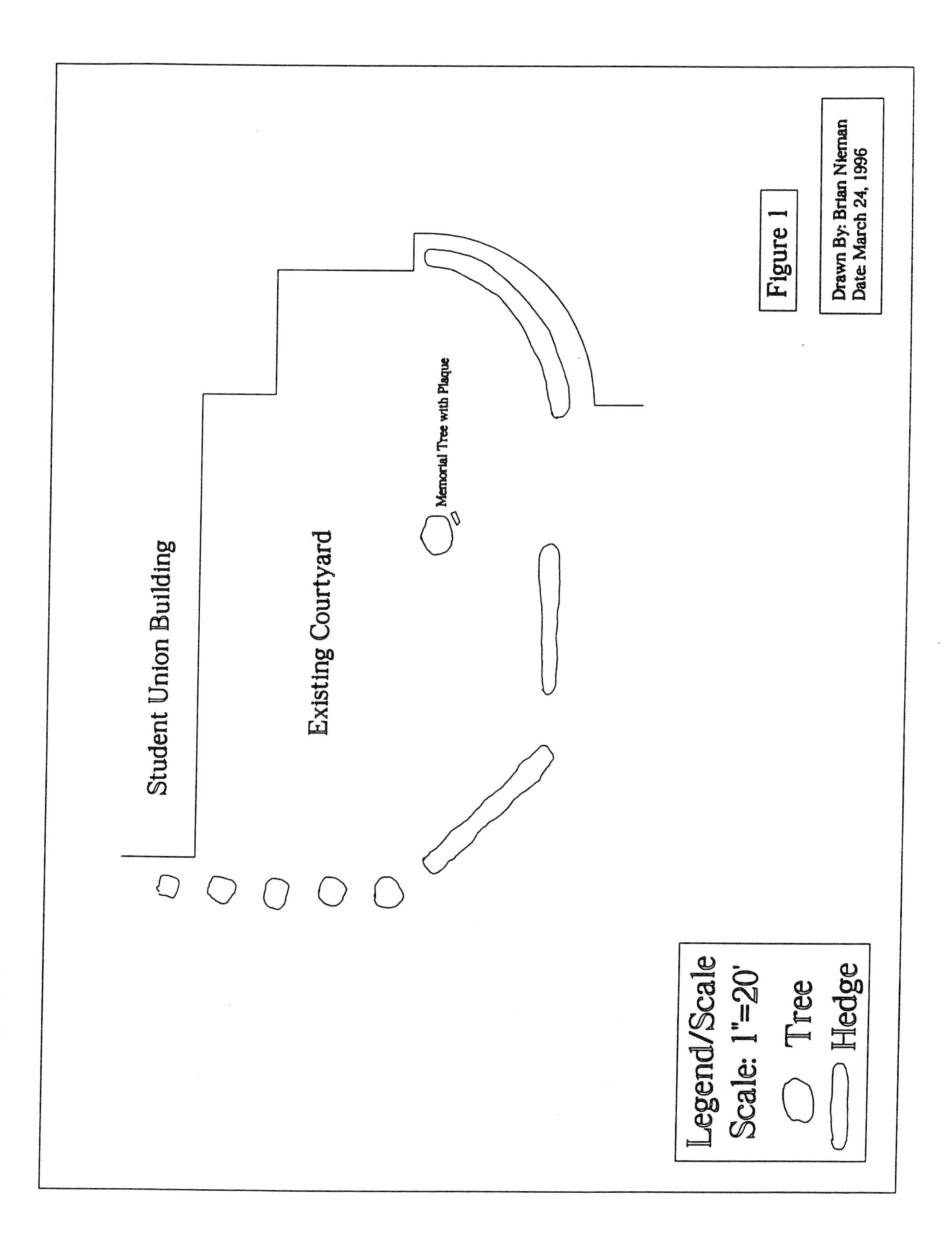
Student Union Building
Existing Courtyard
Memorial Tree with Plaque
Figure 1
Drawn By: Brian Nieman
Date: March 24, 1996
Legend/Scale
Scale: 1"=20'
Tree
Hedge

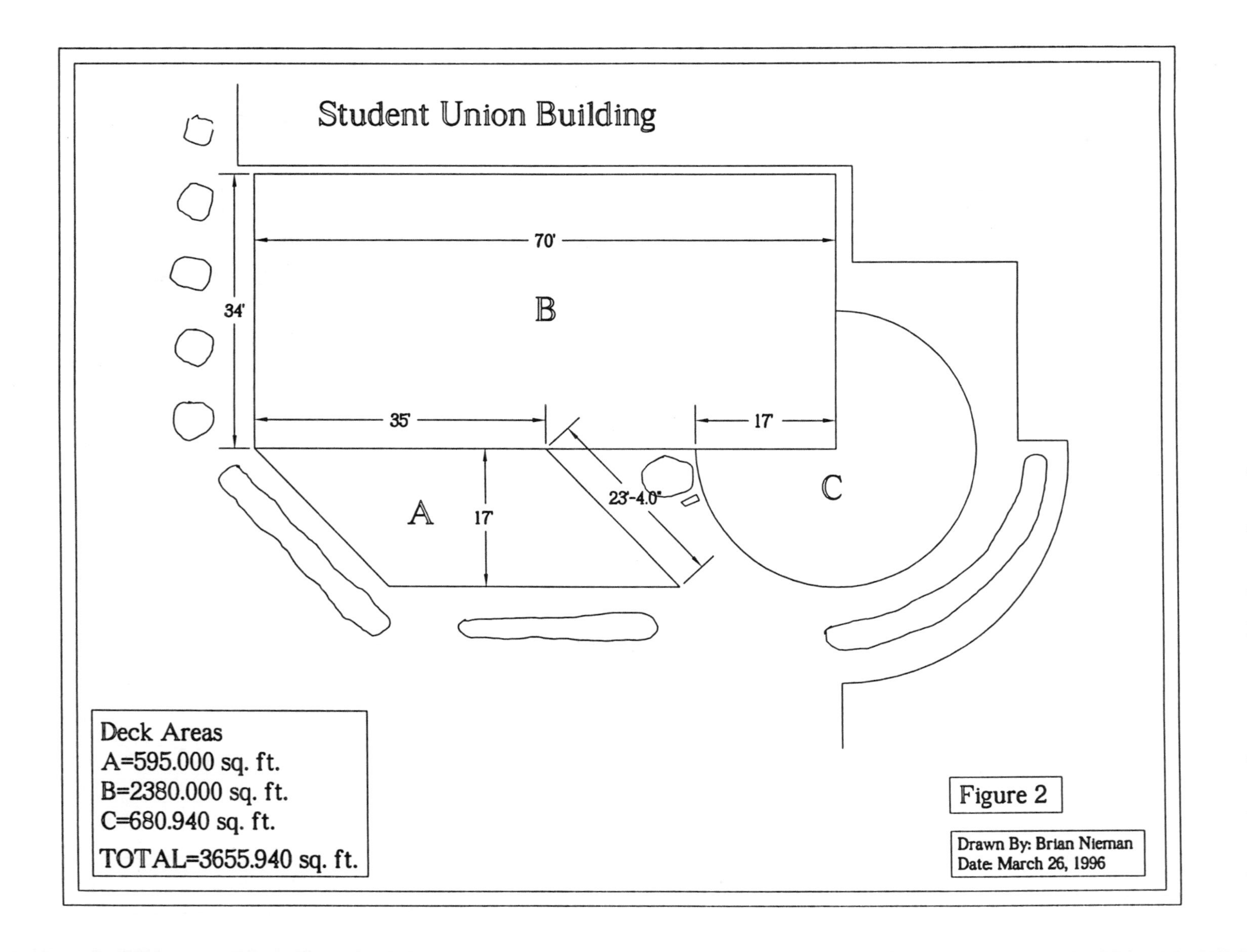
Student Union Building
70'
B
34'
35'
17'
A
17'
23'-4.0"
C
Deck Areas
A=595.000 sq. ft.
B=2380.000 sq. ft.
C=680.940 sq. ft.
TOTAL=3655.940 sq. ft.
Figure 2
Drawn By: Brian Nieman
Date: March 26, 1996

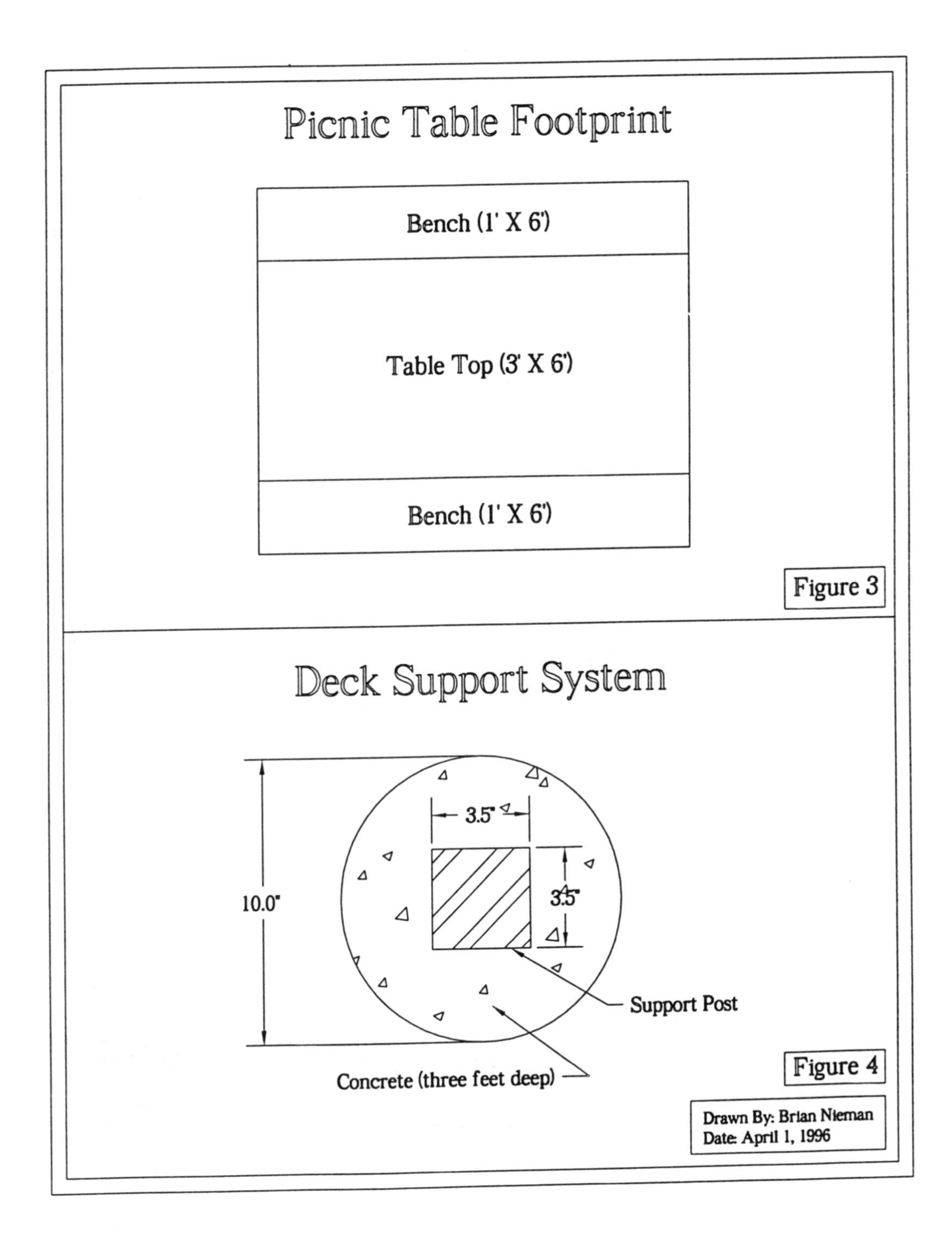
Picnic Table Footprint
Bench (1' X 6')
Table Top (3' X 6')
Bench (1' X 6')
Figure 3
Deck Support System
3.5"
3.5"
10.0"
Support Post
Concrete (three feet deep)
Figure 4
Drawn By: Brian Nieman
Date: April 1, 1996

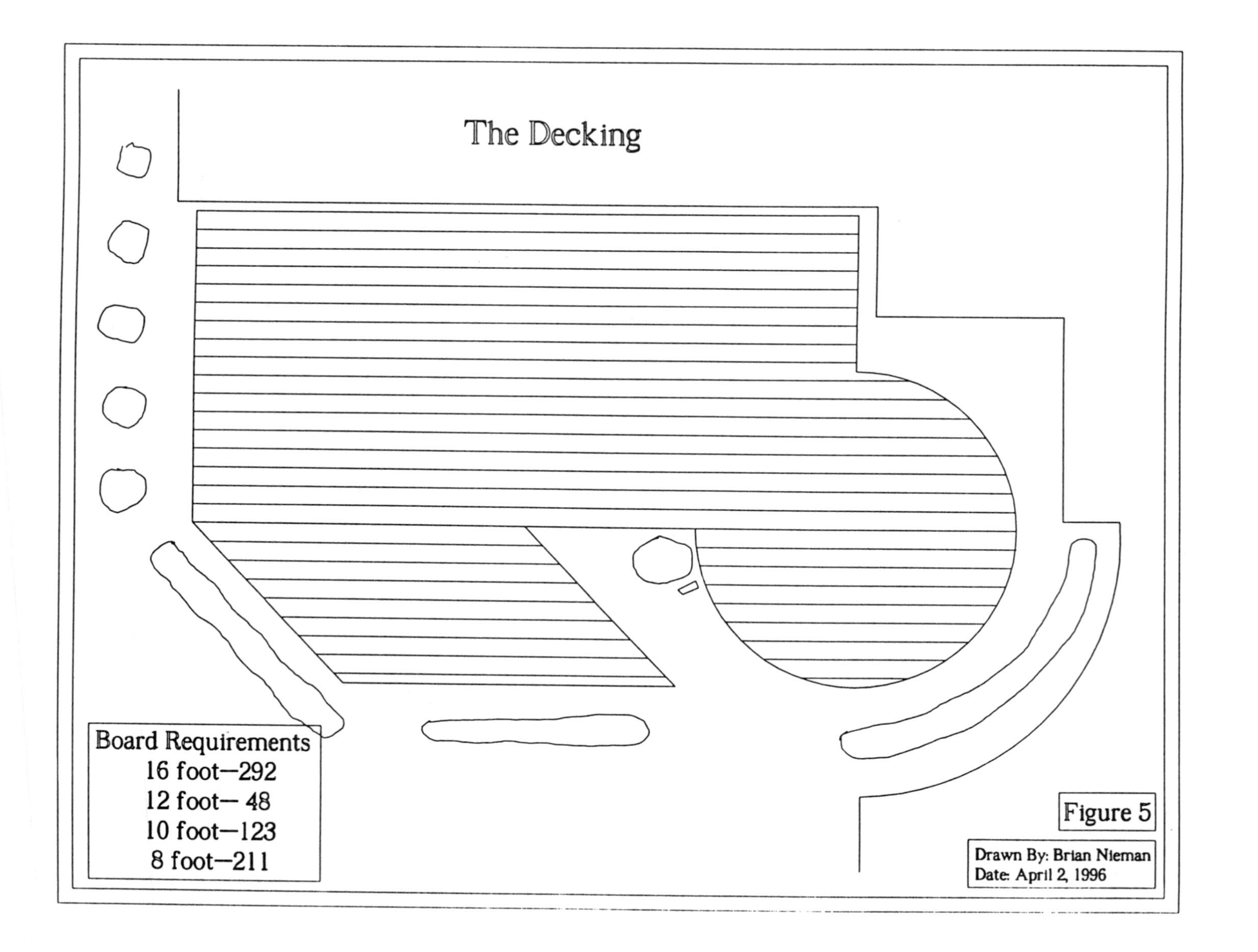
The Decking
Board Requirements
16 foot—292
12 foot— 48
10 foot—123
8 foot—211
Figure 5
Drawn By: Brian Nieman
Date: April 2, 1996

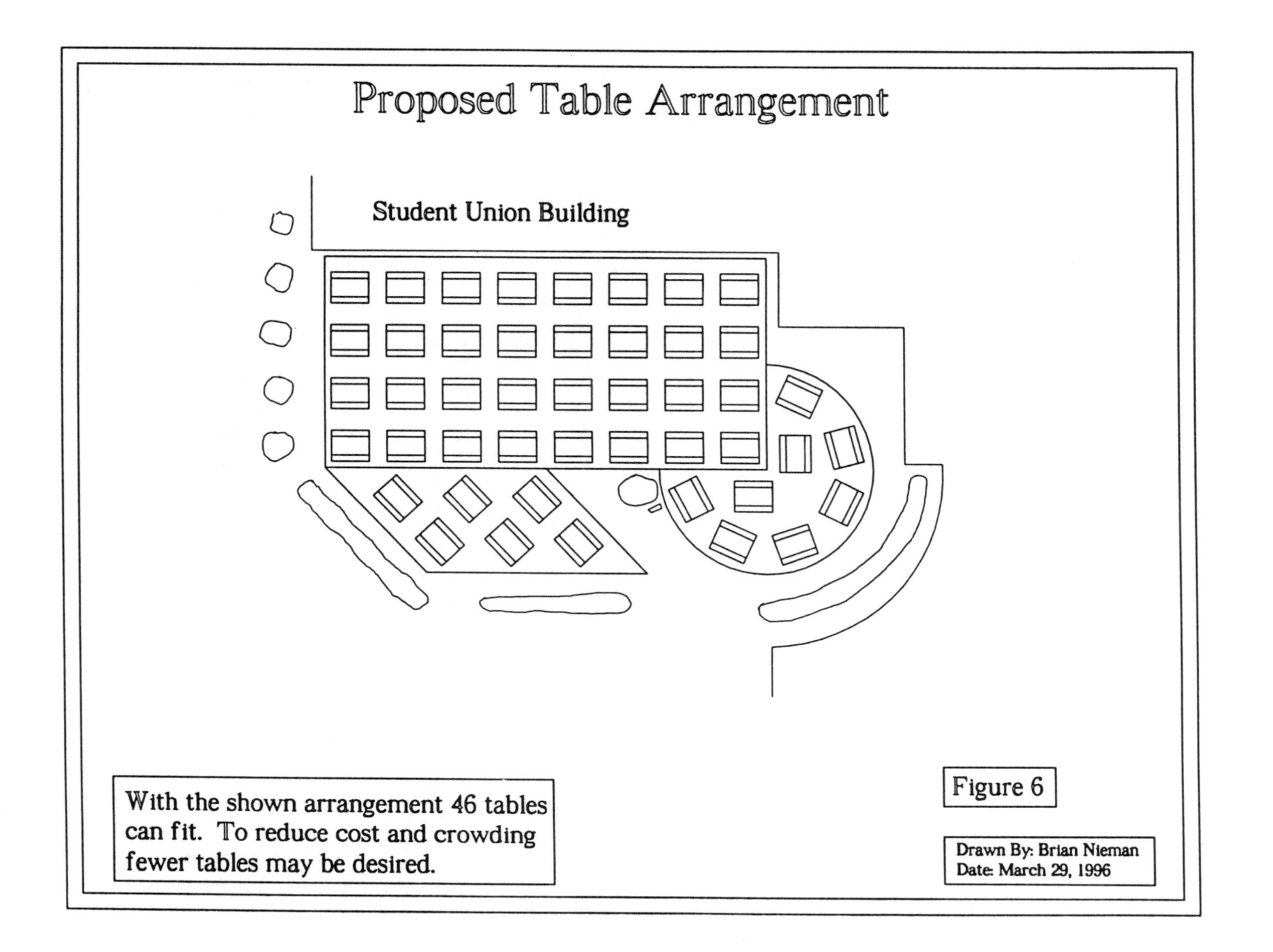
Proposed Table Arrangement
Student Union Building
With the shown arrangement 46 tables can fit. To reduce cost and crowding fewer tables may be desired.
Figure 6
Drawn By: Brian Nieman
Date: March 29, 1996

Title: Decked Out

Notes for the Instructor

This project is designed to be used as a capstone activity for a block of instruction in elementary algebra. It can be handed out at the end of the block and worked on intensively; in this mode, the project should take an average student about 4 hours to complete all requirements and to prepare a written report. However, we generally prefer to hand out projects such as this on the first day of the block, and to have a brief introduction and statement of the problem by an architectural, civil engineering, or other construction-oriented instructor (either in person when enrollment is small or on videotape when enrollment is large). The thrust of this introduction is that there's a construction job to be done and that doing the job is the easy part; the hard part is planning it out to meet all constraints at minimal cost, and that we're counting on the students to do this for us. The students are not sure how to do this yet (on the first lesson of the block), but, by learning the algebraic material in the block they are beginning, they will have the tools they need to do the project. The architect/engineer goes away, and we spend the block learning algebra. As we cover skills that are used on the project, we discuss the connection and encourage the students to work on the project requirements in parallel, and to ask questions that arise along the way. On the second-to-last lesson of the block, the architect/engineer returns, project teams brief their solutions and hand in their reports, and the architect/engineer discusses extensions to the project and how learning algebra will pay off for those students who have chosen or will choose architectural design or engineering when they enter their program.

This is a realistic problem; you may be able to change the requirements or the geometry to match a well-recognized building at your location, or relate it to another piece of construction that has recently been done at your institution. Students find it fascinating that they can use some simple ideas to save money on projects that architects and engineers routinely do, and to see mathematics in projects all around them that they had before just taken for granted.

This is a good project for challenging student intuition, and then demonstrating that intuition (though important) can be unreliable and not acceptable as a substitute for analysis. Survey the students at the beginning of the project about which concrete contractor they think they should use, what pattern to use for the decking, etc. You will get a large distribution of answers. Then ask how they think engineers solve such problems; many will guess that they just wing it based on intuition. Revisit these questions during the project briefings, and cost out some of the original student proposals; the role of analysis is now clear, and the rewards (financial, and maybe aesthetic) are tangible. Students appreciate the need for algebraic analysis, especially when they get to pocket the difference.

A point you can make is that there are really two kinds of optimization. We can have all the facts available and be able to manipulate them to an absolute decision, such as which contractor to call for the amount of concrete which we will require. Then there is the constrained optimization problem, where we may have all the information but it is just too difficult to manipulate the facts down to an absolute decision (such as, "lay the deck boards like this to absolutely minimize the amount of waste). This demonstrates that, even with quantitative analysis, there is room (and there is necessity) for some judgment in satisfying the constraints while still seeking a solution that we hope is at least close to optimal.

Interdisciplinary Lively Application Project

Title: Parachute Panic

Authors: Debra Schnelle
Barbra S. Melendez

Department of Mathematical Sciences and Department of Physics, United States Military Academy, West Point, New York

Editors: David C. Arney
Kathleen G. Snook

Mathematics Classifications: Differential Equations, Calculus

Disciplinary Classifications: Engineering, Physics

Prerequisite Skills:

1. Modeling with differential equations
2. Solving Second-Order differential equations
3. Vector Algebra
4. Vector Calculus (Differentiation of Vector Functions)
5. Projectile Motion and Parametric Equations

Physical Concepts Examined:

1. Newton's Second Law of Motion
2. Conservation of Momentum

Materials Available:

1. Problem Statement (4 Parts); Student
2. Sample Solution (4 Parts); Instructor
3. Notes for the Instructor

Computing Requirements: (none)

INTRODUCTION

After taking the ground preparation portion of your free-fall parachute lessons, you are ready for the airborne preparation portion and your first parachute jump. After a long, exhausting day of training, you fall asleep dreaming about the activities of the next day. You dream about the following scenario:

> You are in the airplane waiting for further instructions when suddenly the pilot of the aircraft hands you a pre-packed parachute. He tells you the airplane is suffering mechanical difficulties and everyone must jump immediately. You put on the parachute as you have been taught. Your free-fall instructor opens the exit door, says "good luck," and proceeds to push you out the door. As you are about to be pushed out the open door you notice that the aircraft is flying over level ground, and it appears the aircraft is flying horizontally at a constant altitude. When pushed, you come straight out from the side of the aircraft and then begin to fall to the earth.
>
> Not being sure when to pull your parachute cord, and recalling the television shows you have seen where people float for a while before opening their parachutes, you decide to wait for a period of time before pulling the parachute cord. Realizing, after a short period, that you feel more like you are speeding toward the earth than floating, you pull the parachute cord and the parachute opens immediately. You arrive safely on the ground, but realize that you could have been seriously injured.

This realization wakes you from your dream and motivates you to study the dynamics of parachute jumping in a bit more detail. In particular, with a little research about the parachute jump in your dream, you establish the following scenario and requirements to analyze to become more familiar with the issues in parachute jumping.

THE FALL.

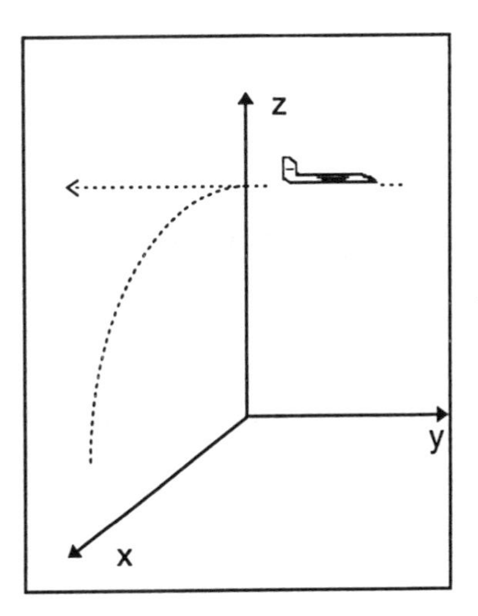

At the local airport's free fall parachute jump school, a student whose mass is 105 kg, leaps with a velocity of .555 m/s straight out (x-axis) from the side of an airplane moving at 115 m/s (y-axis). The airplane is flying at an altitude of 4,000 meters.

When the jump is made, he is high enough in altitude that the air resistance can be considered negligible, and he falls as a freely falling body for 11.5 seconds until he pulls the parachute cord. After that point, the fall becomes a three-dimensional projectile motion problem with non-constant acceleration along each axis as air resistance must now be taken into account.

When the parachute cord is pulled, assume that the parachute deploys immediately and exerts a combined drag and air resistance force of magnitude $D = dv^2$, where d is the drag coefficient due to the parachute and air resistance, and v is the velocity of the student. In this case, we will let $d = 20$, but normally the value of d is difficult to determine, since it depends on so many variables. Assume the drag force is exerted only along the vertical direction (z-axis), and that the effect of air resistance along the y-direction can be modeled as a force of magnitude $\mathbf{F} = -b\mathbf{v}$, where the bold type indicates vectors and the minus sign indicates that the force opposes the direction of motion and b is the drag coefficient due to air resistance. In this case, we will let $b = 10$.

Simultaneous to the opening of the parachute, the student experiences a crosswind in the positive x-direction of 1.2 m/s. The force due to this crosswind can be expressed as $\mathbf{G} = b(\mathbf{w} - \mathbf{v})$, where b is the drag coefficient due to air resistance, $\mathbf{v}$ is the speed of the student, and $\mathbf{w}$ is the cross wind speed.

THE IMPACT.

When the student lands, the impact with the ground can be considered as a collision between the student and the earth.

In the majority of fatal collisions, head injury is responsible for death. Rapid deceleration of the head, even without fracture, can be fatal due to the shear strain on the brain stem (shear means that a differential force has been applied over the surface of an object, thus changing its shape). A measure of this shear strain to the head is the severity index, I:

$$I = \left[\frac{2v}{g\Delta t}\right]^{2.5}(\Delta t)$$

where $\boldsymbol{v}$ is the velocity at impact, Δt is the duration of impact, and g is the acceleration due to gravity. When the severity index for a collision is above 1,000, the collision is fatal. When the value of the severity index is approximately 400, unconsciousness and mild concussion are the result.

The stress on the long bones of the legs during the collision is compressive. When the compression force per unit area exceeds the ultimate tensile strength of the bone (given below for various bones, with their associated cross-sectional areas), the bone breaks in compression.

Bone	**Ultimate Tensile Strength (N/m^2)**	**Average Cross-Sectional Area (m^2)**
Femur	1.21×10^8	5.81×10^{-4}
Tibia	1.40×10^8	3.23×10^{-4}
Spinal Cord (back)	2.20×10^8	4.42×10^{-4}
Spinal Cord (neck)	1.80×10^8	4.42×10^{-4}

The compressive force, ***F***, experienced as a result of the impact on the earth (or water) can be determined from Newton's 2d law, written in terms of the quantity momentum. Momentum, ***p***, is defined as ***p*** = *m**v*** and Newton's 2d law in terms of this quantity states:

$$F = \frac{\Delta p}{\Delta t}$$

where $\Delta \boldsymbol{p}$ is the change in momentum and $\Delta \boldsymbol{t}$ is the time duration of the collision, or impact between the student and the earth (or water).

If the student is fortunate enough to land in a lake, assume that he enters the water at an angle in the safest possible way: such that his pointed toes touch the water first, arms are above his head and clasped together, and he doesn't bend forward.

If the student lands on the ground, assume that he performs a perfect "parachute landing fall," as he was taught in his free-fall class. The time of his impact will depend upon how soft the ground is where he lands. Snow or soft sand, for example, would significantly extend his duration of impact and thus reduce the force of impact.

You know the initial velocities, along each of the coordinate axes, of the student as he exited the airplane. In the first two parts, you will be determining the time of the student's fall to the earth and the position coordinates of his landing.

Part 1.

Determine how long it takes the student to fall to the earth. As a first step, you need to think of the assumptions you need to make to begin the modeling process. *Hint*: Determine the time of the fall from the motion along the vertical (z-axis) direction. Then, since projectile motion is the vector sum of three independent one-dimensional motion events all linked by a common time, the time of the fall along the vertical direction will be the time of the fall along the other two dimensions as well.

A. For the first 11.5 seconds, the student is in free fall and his motion can be described by the kinematics equation. (Remember, during this time period the air is sufficiently thin so that air resistance can be ignored.) Determine the coordinates (x,y,z) at which he pulls the parachute cord.

B. Determine his velocity in each direction (v_x, v_y, v_z), when he pulls the parachute cord.

C. Now he has pulled the parachute cord, and the kinematics equations can no longer be used since his acceleration is no longer constant. Draw a free-body diagram of the force acting upon the student. *Hint*: Draw two free-body diagrams; one for the z-axis and one for the xy plane so that the forces may be clearly identified and drawn.

D. Determine the terminal velocity of the man-parachute system (in the vertical direction only). This will be the velocity at which the drag force of the parachute is equal and opposite to the force of gravity upon the student. Recall that the drag force is $\boldsymbol{D} = dv^2$.

E. The student is monitoring his rate of descent. Six seconds after his parachute opens he feels that he is no longer speeding up (accelerating), but that his speed now seems constant. He checks his wrist altimeter and notes his altitude is 3,298 meters. He continues to fall at this constant velocity until he collides with the ground. How long did he fall after reaching terminal velocity?

F. Find the total time of the fall.

Part 2.

Determine the position coordinates of his landing.

A. Apply Newton's 2d Law to the student's motion in the horizontal (x and y) directions and develop the second-order differential equations that describe his motion in these directions. Motion along the y-axis can be modeled by the homogeneous second-order differential equation $my'' + by' = 0$, and motion along the x-axis can be modeled by the nonhomogeneous second-order differential equation $mx'' + bx' = bw$, where m is the mass of the student, b is the coefficient of air resistance, and w is the wind speed.

B. Solve these differential equations for the position function, where the initial conditions are the position and velocity values at the time the parachute was opened.

C. Find the position coordinates of his landing (x, y, z).

Part 3.

In this requirement, you will determine the student's final velocity (i.e., his velocity immediately before impact) and the force of the collision upon the student.

A. Write the equations describing the student's velocity in the x and y directions as a function of time.

B. Write an expression for the student's final velocity in vector notation.

C. What is his speed at this time?

D. Calculate the change in momentum he experiences as a result of his collision with the ground. Recall that the change in momentum, $\boldsymbol{p}$, is given by the expression $\boldsymbol{p}_{\text{final}} - \boldsymbol{p}_{\text{initial}}$. *Hint*: Consider what the student's momentum must be *after* the collision. Also, make the simplifying assumption that the collision takes place along a single direction; that is, no rotation, bending, or twisting upon impact will occur. Thus, the final direction of motion will be the only direction along which the collision will take place.

E. What is the force of impact as a result of this collision? Recall from earlier work that $\boldsymbol{F} = \Delta\boldsymbol{p}/\Delta\boldsymbol{t}$, where $\Delta\boldsymbol{t}$ is the time duration of the collision. Assume the student landed in thick bushes and his time of impact was 200 milliseconds.

Part 4. What injuries (if any) did the student sustain upon landing?

A. What bones were broken, if any?

B. What is the severity index for an impact at this velocity?

C. In conclusion, did the student survive? If so, what were his major injuries as a result of this experience.

Title: Parachute Panic

SAMPLE SOLUTION

Possible Simplifying Assumptions (Instructors may wish to provide some of these to their students to help begin the modeling process).

A. We are not considering the complexities of landing such as landing on uneven ground or in the trees.

B. We assume the parachute opens instantaneously. The drag force $D = dv^2$ takes effect immediately with no period of transition.

C. The student does not bend or twist upon landing, and thus the breaking torques of various bones and ligaments are not being considered.

D. The student leaves the airplane by jumping straight out with no rotation.

E. Biomechanics Data: The information given for the severity index and the bone compression values for the tibia and femur are correct (although approximate). The spinal fracture data is fictitious. The spine is much more susceptible to shearing and twisting forces than compression forces, and in this type of fall would almost certainly suffer an injury due to those motions than to compression upon landing. The cross-sectional areas for all bones except the tibia are also fictitious.

F. Note that the muscle forces needed to provide the deceleration produce great tensile stresses on ligaments and tendons. Additionally, the failure level for these are much lower than for bone.

G. Assume that the impact force is transmitted without attenuation throughout the body.

Part 1. Time of the Fall.

A. The student waits 11.5 seconds before pulling his parachute open. During this period of time, he is in free fall (ignoring air resistance). His initial velocity in the vertical direction (z-axis) is zero, and the acceleration is due only to the gravitational force and is equal to -9.80 m/s^2. Thus, the student's vertical position at the time he deploys his parachute is given by the equation

$$z = z_0 + v_{z0}t + 0.5\, a_z t^2$$

$$z = 4000 + 0 + 0.5\,(-9.80)(11.5)^2$$

$$\mathbf{z = 3{,}352\ m}$$

His initial velocity in the *x*-direction is 0.555 m/s, and there is no acceleration in this direction (air resistance is negligible during the first 11.5 seconds). So, his position in the *x*-direction is given by

$$x = x_0 + v_{x0}t + 0.5\, a_x t^2$$

$$x = 0 + 0.555(11.5) + 0$$

$$\mathbf{x = 6.4\ m}$$

Similarly, the initial velocity in the *y*-direction is 115 m/s, and the acceleration in this direction is zero. Thus, the position coordinate in the *y*-direction is given by:

$$y = y_0 + v_{y0}t + 0.5\, a_y t^2$$

$$y = 0 + 115(11.5) + 0$$

$$\mathbf{y = 1{,}322.5\ m}$$

So, the position coordinate at which the student opened his parachute is **(x, y, z) = (6.4, 1322.5, 3352) m**.

B. When the student pulls the parachute cord, his velocity in the *x*- and *y*-directions will be the same as the initial values for those velocities when he leaped from the aircraft. His vertical velocity can be determined by

$$v_z = v_{z0} + a_z t$$

$$v_z = 0 + (-9.8)(11.5)$$

$$\mathbf{v_z = -112.7\ m/s}$$

So, the velocity of the student when he pulls the parachute is given by **(v_x, v_y, v_z) = (0.555, 115, -112.7) m/s.**

C. The forces acting upon the student after the parachute is opened can be depicted as below on a free-body diagram.

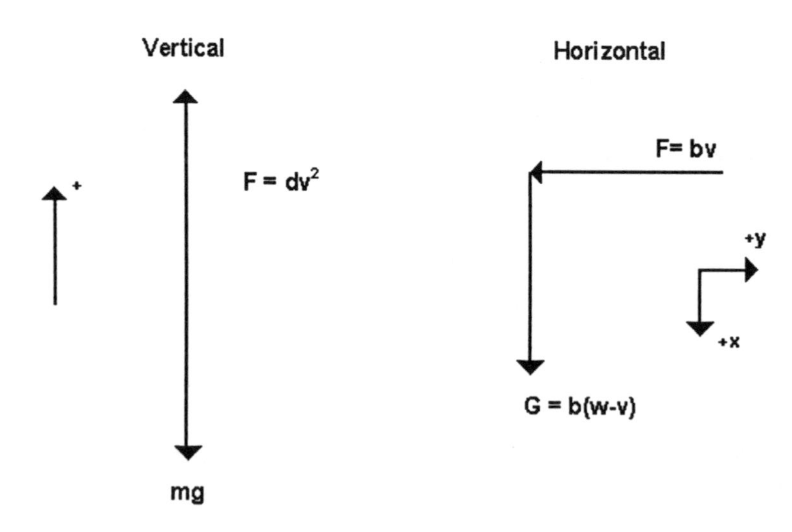

D. Now we can determine the terminal velocity of the student. He will reach terminal velocity when the drag force of the parachute, $\boldsymbol{D = dv^2}$, is equal to his weight, ***mg***, so that his resultant acceleration is zero -- in other words, when $\boldsymbol{dv^2 + mg = 0}$.

$$dv^2 = -mg$$

$$v = \sqrt{\frac{-mg}{d}}$$

$$v = \sqrt{\frac{-(105)(-9.8)}{20}}$$

$v = 7.1042$ in the downward direction

Therefore, $\boldsymbol{v = -7.10\,k}$ m/s

E. Since at terminal velocity the student is falling downward at a constant rate of 7.1042 m/s, the time required to fall the remaining distance of 3,298 meters is:

$$(3298)/(7.1042) = 464 \text{ seconds}$$

F. The total time of the fall is calculated by summing the time spent in free fall motion and the time with the parachute open, both before and after terminal velocity was reached.

$$\text{Total Time} = 11.5 + 6.0 + 464$$

$$\textbf{Total Time = 481.7 seconds}$$

Part 2. Determining the landing coordinates.

A. Applying Newton's 2d Law to the student's motion in both the *x*- and *y*-directions, we obtain the following equations of motion.

***x*-direction:** $\mathbf{F_x = ma_x}$

where $F = b(w - v)$

so,

$$b(w-v) = ma_x$$
$$bw - bv = ma$$
$$ma + bv = bw$$
$$mx'' + bx' = bw$$
$$\mathbf{103x'' + 10x' = 12}$$

***y*-direction:** $\mathbf{F_y = ma_y}$

where $F = -bv$

so,

$$-bv = ma_y$$
$$ma + bv = 0$$
$$my'' + by' = 0$$
$$\mathbf{103y'' + 10y' = 0}$$

where *m* is the mass of the student, *b* is the coefficient of air resistance, and *w* is the wind speed.

B. We can solve these equations for the position function. The initial conditions are the position and the velocity values *at the time the parachute was opened.*

***y*-direction**: Use the characteristic equation $mr^2 + br = 0$.

$$r = 0 \text{ and } r = -(b/m)$$

so, the characteristic solution is:

$$y(t) = C_1 e^0 + C_2 e^{-(bt/m)}$$

where the constants C_1 and C_2 are found by using the initial conditions.

Note that these are the conditions at time $t = 0$, that is, the time at which the parachute is opened and air resistance could no longer be considered as negligible. So at this time, $y = 1{,}322$ m and $v_y = 115$ m/s (from **Part 1.A.** above).

$$1322.5 = C_1 + C_2 \quad \text{and} \quad 115 = -(b/m)\, C_2$$

where $b = 10$ kg/m and $m = 103$ *kg.*

Therefore, $C_2 = -1184.5$ $\quad C_1 = 2507$

And we see that: $\mathbf{y(t) = 2507 - 1184.5\, e^{-(10/103)t}}$

***x*-direction**: In the *x*-direction, the differential equation is non-homogeneous, so the associated equation $mx'' + bx' = 0$ should be solved first, and then the non-homogeneous equation $mx'' + bx' = bw$ can be solved. The characteristic equation for the homogeneous differential equation is $mr^2 + br = 0$ where the solutions to this equation are:

$$r = 0 \quad \text{and} \quad r = -(b/m)$$

So, the complementary solution is:

$$x_c = C_1 e^0 + C_2 e^{-(bt/m)}$$

Now solve for the particular solution, x_p, of the non-homogeneous equation $mx'' + bx' = bw$. Assume $x_p = At$ (an initial guess of $x_p = A$ would lead to repeated roots of 0). Then, $x_p' = A$, and $x_p'' = 0$. Substituting this solution into the differential equation yields the following analysis:

$$m(0) + bA = bw$$

$$A = w \text{ and } \quad x_p = wt$$

The general solution will be $x = x_c + x_p$ The general solution for our non-homogeneous differential equation is:

$$x = C_1 + C_2\, e^{-(bt/m)} + wt$$

Therefore, $\mathbf{x(t) = C_1 + C_2\, e^{-(10/103)t} + 1.2\, t}$

Now we can use the initial condition to determine the constants C_1 and C_2. Note that again these are the conditions at time $t = 0$, that is, the time at which the parachute opened and air resistance could no longer be considered as negligible. So, at this time, x = 6.3825 m and v_x = 0.555 m/s (from **Part 1.A.** above).

$$6.3825 = C_1 + C_2 \quad \text{and} \quad 0.555 = -(b/m)\, C_2 + w$$

so, $C_2 = 6.6435$ and $C_1 = -0.261$

And we see that: $\mathbf{x(t) = -0.261 + 6.6435\, e^{-(10/103)t} + 1.2\, t}$

C. In order to find the position coordinates of the student's landing site, we use the position equations and the total time of the parachute phase of the jump. Given the position equations found above, simply substitute into these equations the value of the time when the student lands (the total time experienced by the man-parachute system), t = 470.2324 seconds.

We find the position is: **x = 1,155.5554 m, y = 2,507 m,** and **z = 0 m.**

Part 3. Now we want to determine the student's final velocity (i.e., his velocity immediately before impact) and the force of the collision upon the student.

A. The velocity equations as a function of time will be the derivatives, with respect to time, of the position equations. Therefore,

***y*-direction:** position equation ---> $y(t) = 2507 - 1184.5\ e^{-(10/103)t}$

velocity equation ---> $v_y(t) = (-\ 1184.5)(-10/103)\ e^{-(10/103)t}$

$= 115\ \ e^{-(10/103)t}$

***x*-direction:** position equation -> $x(t) = -\ 0.261 + 6.6435\ e^{-(10/103)t} + 1.2\ t$

velocity equation -> $v_x(t) = (6.6435)(-10/103)\ \ e^{-(10/103)t} + 1.2$

$= \ 0.645\ \ e^{-(10/103)t} + 1.2$

B. At the time of impact, the student will have reached terminal velocity in the *z*-direction; therefore, we know $v_z(t) = -\ 7.1042$ m/s. Since we determined the total time of the fall is 481.7324 seconds, and he opened his parachute at 11.5 seconds, we are interested in his velocity at 470.2324 seconds. This is the time value we substitute into the velocity equations above to obtain the velocity upon impact in each direction.

$v_x(470.2324) = \ 0.645\ \ e^{-(10/103)(470.2324)} + 1.2 \approx 1.20$

$v_y(470.2324) = 115\ \ e^{-(10/103)(470.2324)} \approx 0.00$

$v_z(470.2324) = -\ 7.1042$ m/s

In vector notation we can write, $\boldsymbol{v} = v_x\mathbf{i} + v_y\mathbf{j} + v_z\mathbf{k}$

$\boldsymbol{v} = (1.20\ \mathbf{i} + 0\ \mathbf{j} - 7.10\ \mathbf{k}\ )$ m/s

C. The magnitude of the velocity vector gives the speed of the student at the time of impact. In order to determine the speed we can use the Pythagorean theorem.

$$\text{speed} = \sqrt{v_x^2 + v_y^2 + v_z^2}$$

$$= \sqrt{1.2^2 + 0^2 + (-7.1042)^2}$$

$$= \mathbf{7.20\ m/s}$$

D. When the student collides with the ground, he experiences a change in momentum, $\boldsymbol{p}$ given by $\boldsymbol{p}_{\text{final}} - \boldsymbol{p}_{\text{initial}}$. After the student collides with the ground, he will have no momentum (as he lies on the ground assessing his possible injuries). Therefore, $\boldsymbol{p}_{\text{final}} = 0$. Additionally, if we assume that the collision takes place along a single direction we can use the speed on impact to determine $\boldsymbol{p}_{\text{initial}}$. Momentum is given by $m\boldsymbol{v}$, where m is the mass of the student and $\boldsymbol{v}$ is his velocity upon impact. We can then determine his initial momentum as:

$$\boldsymbol{p}_{\text{initial}} = m\boldsymbol{v} = (103)(7.2048) = 742.09 \text{ kg·m/s}$$

And since his final momentum is 0, the change in momentum is 0 - 742.09 = - 742.0944 kg·m/s. This can also be described as 742.09 N·s, where N = 1 kg·m/s^2 indicates force in units of Newtons.

E. We can determine the magnitude of the force of impact using the equation $|F| = |\Delta p / \Delta t|$. Since the student landed in thick bushes, his time of impact was 200 milliseconds (0.200 seconds). Therefore,

$$|F| = (742.0944) / (.200) = \mathbf{3{,}710\ N}$$

Part 4. With the information we have found so far we can determine whether the student sustained any injuries.

A. Table 1 gives us the ultimate tensile strength in N/m^2 and the average cross-sectional area in m^2 for four bones vulnerable to injury in this type of fall. By multiplying the tensile strength and the cross section, we obtain the maximum force each bone is able to sustain without breaking. The following table provides this additional information for analysis.

Bone	Ultimate Tensile Strength (N/m^2)	Average Cross-Sectional Area (m^2)	Max Sustainable Force (N)
Femur	1.21×10^8	5.81×10^{-4}	70,301
Tibia	1.40×10^8	3.23×10^{-4}	45,220
Spinal Cord (back)	2.20×10^8	4.42×10^{-4}	97,240
Spinal Cord (neck)	1.80×10^8	4.42×10^{-4}	79,560

From Table 2 we see that a force of 3,710.47 N would not break any of the four bones analyzed.

B. Since no major bones were broken, we should check for injury to the head by shear strain. We use the severity index, *I*, to determine possible injury.

$$I = \left[\frac{2v}{g\Delta t}\right]^{2.5}(\Delta t)$$

$$I = \left[\frac{2(7.2048)}{9.8(.200)}\right]^{2.5}(.200)$$

$$\mathbf{I = 29.31}$$

C. The severity injury of 29.31 is not close to the fatal level of 1,000. Therefore, there were no serious injuries to the head. We saw above there were no serious injuries to any of the major bones. Although there may have been no serious injuries, there are a multitude of less serious injuries possible from this type of landing.

Title: Parachute Panic

Notes for the Instructor

The Parachute Panic ILAP is best suited for use in a vector calculus course. Students should have studied or be currently studying elementary differential equations. As a motion analysis problem, this project's application comes from physics and engineering.

This ILAP is based on a comprehensive motion problem. A parachute jumper exits a plane and descends to the ground in two phases; the first with a constant acceleration, and the second with a nonconstant acceleration. The project involves the concepts of Newton's Second Law of Motion and Conservation of Momentum. The solution process includes modeling and solving second-order homogeneous and nonhomogeneous differential equations, as well as using vector algebra, vector calculus, and parametric equations. The student should realize that motion is continuous in three directions, and that the motion in each direction is linked by the parameter time.

You may wish to provide some of these simplifying assumptions to give your inexperienced students help in beginning the modeling process and in thinking of the different models needed in solving the problem. For more experienced students, this process is probably best left to the students. Of course, this is just one possible set of valid assumptions.

A. We are not considering the complexities of landing such as landing on uneven ground or in the trees.

B. We assume the parachute opens instantaneously. The drag force $D = dv^2$ takes effect immediately with no period of transition.

C. The student does not bend or twist upon landing, and thus the breaking torques of various bones and ligaments are not being considered.

D. The student leaves the airplane by jumping straight out with no rotation.

E. Biomechanics Data: The information given for the severity index and the bone compression values for the tibia and femur are correct (although approximate). The spinal fracture data is fictitious. The spine is much more susceptible to shearing and twisting forces than compression forces, and in this type of fall would almost certainly suffer an injury due to those motions rather than to compression upon landing. The cross-sectional areas for all bones except the tibia are also fictitious.

F. Note that the muscle forces needed to provide the deceleration produce great tensile stresses on ligaments and tendons. Additionally, the failure level for these are much lower than for bone.

G. Assume that the impact force is transmitted without attenuation throughout the body.

Parts 1 and 2. The first two requirements ask the student to determine the total time of the parachutist's descent and the position coordinates of the parachutist's landing. In solving complex problems it is useful to break the problem into smaller more manageable sub-problems. Parts 1 and 2 are divided into sub-requirements that assist the student through the solution process. Students, however, tend to lose sight of the overall objective of finding total time and final position. The total time aspect causes the most common problems. Students must analyze the jump in two phases, and each phase has its own start and finish times. Additionally, each phase also has its own start and finish locations. The "resetting of the clock" leads to confusion for some students.

If your students are prepared to be provided only the general requirements without the guiding sub-requirements, you may adjust your requirement page accordingly. You may also want the students to analyze the problem first and then come to you to ask for information they need. You may then provide them guidance as is done in the sub-requirements. In this way you are monitoring their solution process and keeping them on task.

Parts 3 and 4. Part 3 asks students to determine the force at impact and the change in momentum the parachutist experiences. Part 4 then uses this information to determine whether the parachutist sustained any injuries upon landing. Both force and momentum are vector quantities. Requirement 3 can be completed using the velocity vector and vector analysis, or the magnitude of the velocity vector (speed) and linear analysis. Once the force and the speed at impact are obtained, Part 4 simply analyzes those values using information provided about severity of collisions and bone strength.

Again, the requirements as written provide sub-requirements to guide the student through the solution processes. You may want to adjust the requirements for your students, or have them do some initial analysis first and then provide them guidance.

Interdisciplinary Lively Application Project

Title: Flying With Differential Equations

Authors: Joseph D. Myers
Walter S. Barge
Guy Harris

Department of Mathematical Sciences and Department of Civil and Mechanical Engineering, United States Military Academy, West Point, New York

Editors: David C. Arney and Kathleen G. Snook

Mathematics Classifications: Differential Equations, Calculus

Disciplinary Classification: Mechanical & Aeronautical Engineering,

Prerequisite/Corequisite Skills:

1. Modeling with Differential Equations
2. Solving Constant Coefficient, Nonhomogeneous Differential Equations
3. Numerical Solution of 1st-order Differential Equation (Euler's Method and/or a Runge-Kutta Method)

Physical Concepts Examined:

1. Forced Vibrations
2. Mechanical Resonance
3. Motion under Gravity

Materials Available:

1. Problem Statement (4 Parts); Student
2. Sample Solution (4 Parts); Instructor
3. Notes for the Instructor

Computing Requirements:

1. Tools for iterating a difference equation (derived from Euler or Runge-Kutta Methods), such as a spreadsheet)
2. Numerical differential equations solver (optional)

Background Information

> If you have ever looked out a window while in flight, you have probably observed that wings on an airplane are not perfectly rigid. A reasonable amount of flex or flutter is not only tolerated but necessary to prevent the wing from snapping like a piece of peppermint stick candy. In late 1959 and early 1960 two commercial plane crashes involving a relatively new model of prop-jet occurred, illustrating the destructive effects of large mechanical oscillations. ... After a massive technical investigation, the problem was eventually traced in each case to an outboard engine and engine housing. Roughly, it was determined that when each plane surpassed a critical speed of approximately 400 mph, a propeller and engine began to wobble, causing a gyroscopic force, which could not be quelled or damped by the engine housing. This *external vibrational force* was then transferred to the already oscillating wing. This, in itself, need not have been destructively dangerous since aircraft wings are designed to withstand the stress of unusual and excessive forces. ... But unfortunately, after a short period of time during which the engine wobbled rapidly, *the frequency of the impressed force actually slowed* to a point at which it approached and finally coincided with the maximum frequency of wing flutter. The amplitudes of wing flutter became large enough to snap the wing. (Zill, 1989, p. 219)

In this problem, we will examine the phenomena which caused the airplane wing in the above scenario to snap. This phenomena is called mechanical resonance. Any structure or mechanical system is susceptible to damage by the forces of resonance. The following excerpt presents a clear and interesting discussion of resonance.

> When the frequency of a periodic external force applied to a mechanical system is related in a simple way to the natural frequency of the system, mechanical resonance may occur which builds up the oscillations to such tremendous magnitudes that the system may fall apart. A company of soldiers marching in step across a bridge may in this manner cause the bridge to collapse even though the bridge would have been strong enough to carry many more soldiers had they marched out of step. For this reason soldiers [are] required to

> "break step" [when] crossing a bridge. In an analogous manner, it may be possible for a musical note of proper characteristic frequency to shatter a glass. Because of the great damages which may thus occur, mechanical resonance is in general something which needs to be avoided, especially by the engineer in designing structure or vibrating systems. (Spiegel, 1981, p. 239)

We can use differential equations to construct a simple model of wing flutter. The *center-of-mass* of an object is the point at which we can consider the entire mass of an object to be represented. It is the balance point (in terms of mass) of the object. If we restrict our investigation to the fluttering motion that takes place at the wing's center-of-mass, we can build a simple model of this situation. By restricting our attention to this point, we can then think of the wing as a *spring-mass system.* The *spring* is the wing-body joint that allows the wing's center-of-mass to move up and down in a fluttering motion. The forcing function is the external vibrational force that comes from the wobbling propeller. Remember -- any motion at the wing's center-of-mass will be magnified out at the tip of the wing!

Some Assumptions

For the sake of solving this simple model, we assume that the wing has a mass of m kg and that the wing-body joint will act like a spring with a *spring constant* of k N/m (Newtons per meter). The motion of the wing's center-of-mass is actually a curved arc; however, since we are working with long wingspans, we can simplify the situation by assuming that the center-of-mass moves up and down in a straight line. Assume that, before the propeller begins to wobble, the wing is at rest. Also, assume that any forces which tend to damp the motion of the wing are negligible.

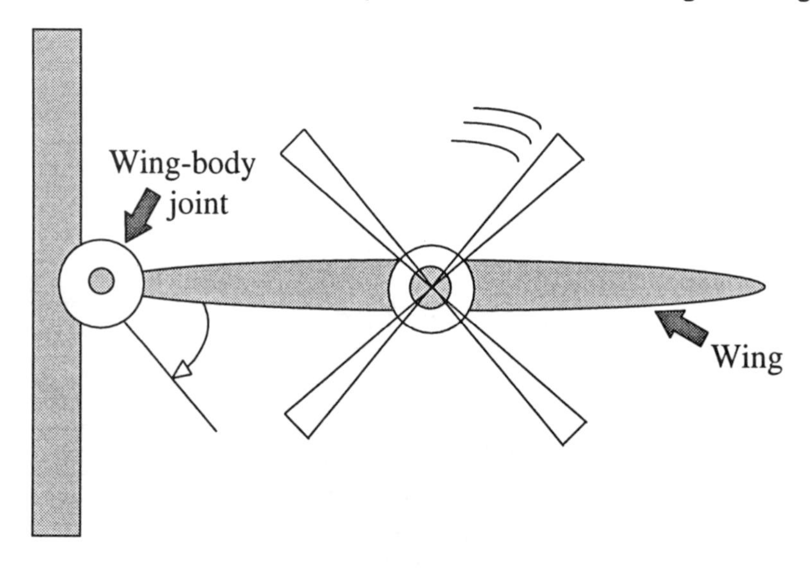

We want to analyze and understand this wing flutter phenomenon, and, as part of our company's design team, we need you to conduct the following step-by-step analysis to determine several of the critical issues of this phenomenon.

Part 1.

Starting at *time* = 0 ($t = 0$), assume the propeller begins to vibrate with a force equal to $F(t) = F_0 \sin wt$.

A. Find the equation of motion for the wing's center-of-mass. Be sure to include the appropriate initial conditions you would need to solve this differential equation.

B. Solve your equation of motion.

C. Now let $m = 850$ kg, k = 13,600 N/m, $w = 8$, and $F_0 = 1{,}550$ N, and write the solution using these values.

D. Plot your displacement function using your calculator or a computer. Include this plot in your submission.

E. Describe the motion of the center-of-mass as t (time) grows large.

Part 2.

Recall from the excerpt above that just before wing failure the frequency of the external force caused by the wobbling propeller actually slowed down. Let's simulate that slowing down by changing our forcing function to $F_2(t) = 1550 \sin 4t$ N.

A. Plot the two forcing functions ($F(t)$ and $F_2(t)$) and include with your submission. What is the principal difference in the two functions?

B. Find the equation of motion using the new forcing function (letting all other conditions remain the same). Again, make sure that you include the appropriate initial conditions. Solve your new equation of motion for the displacement function.

C. Plot this displacement function on your calculator or computer and include with your submission.

D. Describe what happens as t grows large. What consequences does this have for the wing?

Part 3.

Let's now assume that the wing does finally snap. The wing itself is now in free fall towards the earth. Further assume that the wing does not tumble, but instead remains mostly horizontal as it falls. Several forces may be acting upon the wing as it falls. We will consider two: resistive drag force (F_d), and force due to gravity (F_g). (Assume that down is the positive direction.)

The drag force results from collisions between falling objects (in this case the wing) and the air molecules. A common sub-model for drag forces proposes that the force from drag is proportional to some power of the object's velocity. We choose a two-term model here:

$$F_d = -k_1 v - k_2 v^2.$$

(One can think of this as the first two terms of a Taylor Series representation for the drag function.)

The sub-model for force due to gravity should be familiar:

$$F_g = mg.$$

For the purposes of this problem, we will use $k_1 = 10.5$, and $k_2 = 6.5$.

A. Model the velocity of the free-falling wing with a differential equation. Classify the differential equation (order, linearity, homogeneity) and be sure to include the appropriate initial conditions.

B. You notice that this type of differential equation is not as easy to solve using analytic methods. Therefore, you must use a numerical method to approximate the solution. Choose a suitable numerical technique and step size that will allow you to determine the velocity of the wing 4 seconds after it snaps.

C. Assuming its initial vertical velocity is zero, what is the velocity of the wing after it has fallen 50 meters?

D. What is the wing's terminal velocity?

Part 4.

Think critically about the model used in Requirements 1 and 2.

A. What forces are at work on the real plane wing that we did not include in our model. (Hint: Look at the differential equation you used in Requirement 2. What type of motion is being modeled?)

B. What other factors or issues might an engineer want to analyze in his investigation of wing flutter?

C. How would you model flex in the aircraft wing (e.g., with a system of differential equations)?

REFERENCES

Spiegel, M. R. (1981). *Applied differential equations.* Englewood Cliffs, N. J.: Prentice Hall.

Zill, D. G. (1989). *A first course in differential equations with applications.* Boston: PWS-Kent.

Title: Flying With Differential Equations

SAMPLE SOLUTION

Part 1.

A. We can model this situation as one where a periodic external force, $F_0 \sin wt$ (vibration of the propeller), is applied to a spring-mass system (the wing). Newton's Second Law yields:

$$m\,a = \_\,F = m\,y''(t) = -\,k\,y(t) + F_0 \sin wt \qquad (1)$$

Initial conditions: $y(0) = 0,\ y'(0) = 0$

where m = mass of the wing, k = spring constant, w = period of the forcing function, $F_0 = F(0)$, and $y(t)$ = displacement of the wing's center of mass from its resting position.

We can rewrite equation (1) as:

$$m\,y''(t) + k\,y(t) = F_0 \ \sin wt \qquad (2)$$

Initial conditions: $y(0) = 0,\ y'(0) = 0$

B. To find a general solution for equation (2), we look for both the particular and complementary solutions. These are the solutions to the nonhomogeneous and homogeneous problems respectively. First, the complementary solution is found by solving the homogeneous equation:

$$m\,y''(t) + k\,y(t) = 0$$

whose characteristic equation is $r^2 + (k/m) = 0$. The roots of the characteristic equation are $r = \pm i\sqrt{\frac{k}{m}}$.

We find the complementary solution to be:

$$y_c(t) = c_1 \cos \sqrt{\frac{k}{m}}\, t + c_2 \sin \sqrt{\frac{k}{m}}\, t$$

and, since $w_0 = \sqrt{\frac{k}{m}}$, this leads to:

$$y_c(t) = c_1 \cos w_0 t + c_2 \sin w_0 t.$$

Second, looking for a particular solution, we try $y_p(t) = A \sin wt$. Substituting, we obtain:

$$m(- A\, w^2 \sin wt) + k\,(A \sin wt) = F_0 \sin wt\,.$$

So, $$A\,(- mw^2 + k) = F_0$$

and therefore, $$A = F_0 \,/\, (- mw^2 + k) = F_0 \,/\, [m\,(k/m - w^2)]\ .$$

Letting w_0 = the natural period of the spring-mass system = $\sqrt{\frac{k}{m}}$, we see that:

$$A = F_0 \,/\, [m\,(w_0^2 - w^2)]$$

and therefore, $$y_p(t) = [F_0 / [m\,(w_0^2 - w^2)]] \sin wt\ .$$

Therefore, the general solution is the linear combination of the particular and complementary solutions:

$$y(t) = c_1 \cos w_0 t + c_2 \sin w_0 t + [F_0/ [m\,(w_0^2 - w^2)]] \sin wt$$

Using the initial conditions of $y(0) = 0$ and $y'(0) = 0$:

$$y(0) = c_1 + 0 + 0 = 0$$
$$\longrightarrow c_1 = 0$$

$$y'(0) = 0 + w_0\, c_2 + w\, F_0 /[m(w_0^2 - w^2)] = 0$$

$$\longrightarrow c_2 = - (w/w_0) F_0 /[m(w_0^2 - w^2)]$$

Our solution becomes:

$$y(t) = -(w/w_0)[F_0/[m(w_0^2 - w^2)]] \sin w_0 t + [F_0/[m(w_0^2 - w^2)]] \sin wt$$

$$y(t) = [F_0/[m(w_0^2 - w^2)]] [\sin wt - (w/w_0) \sin w_0 t] \qquad \textbf{(3)}$$

C. Letting k = 13,600 N/m, m = 850 kg, w = 8, and F_0 = 1,550 N, we can substitute into equation **(3)** and obtain the equation of motion (don't forget $w_0 = \sqrt{\frac{k}{m}}$).

Since: $w_0 = \sqrt{\frac{k}{m}} = \sqrt{\frac{13{,}600}{850}} = 4$,

$F_0/[m(w_0^2 - w^2)] = -0.03799$ and $w/w_0 = 8/4 = 2$.

Therefore: $y(t) = [-0.03799] [\sin 8t - 2 \sin 4t]$

$$y(t) = 0.07598 \sin 4t - 0.03799 \sin 8t$$

D. Plotting this solution, we obtain the following graph

Graph of $y(t) = 0.07598 \sin 4t - 0.03799 \sin 8t$

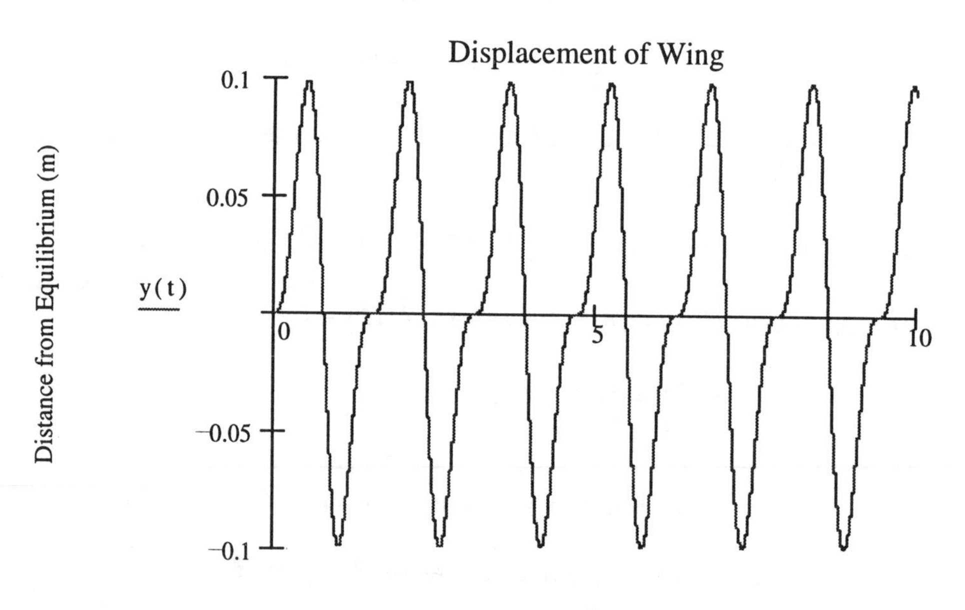

E. As time (t) grows large, we see the center-of-mass of the wing is oscillating on periodic cycle. The maximum and minimum points are equidistant from the axis defined by the wing at rest. Therefore, if the wing is stable through one complete cycle it should remain stable as long as no other forces are applied.

Part 2.

For this requirement, we simulate the slowing down of the propeller by changing the forcing function to $F_2(t) = 1550 \sin 4t$.

A. Plotting the two forcing functions ($F(t)$ and $F_2(t)$) on a calculator or computer, we obtain the following graphs.

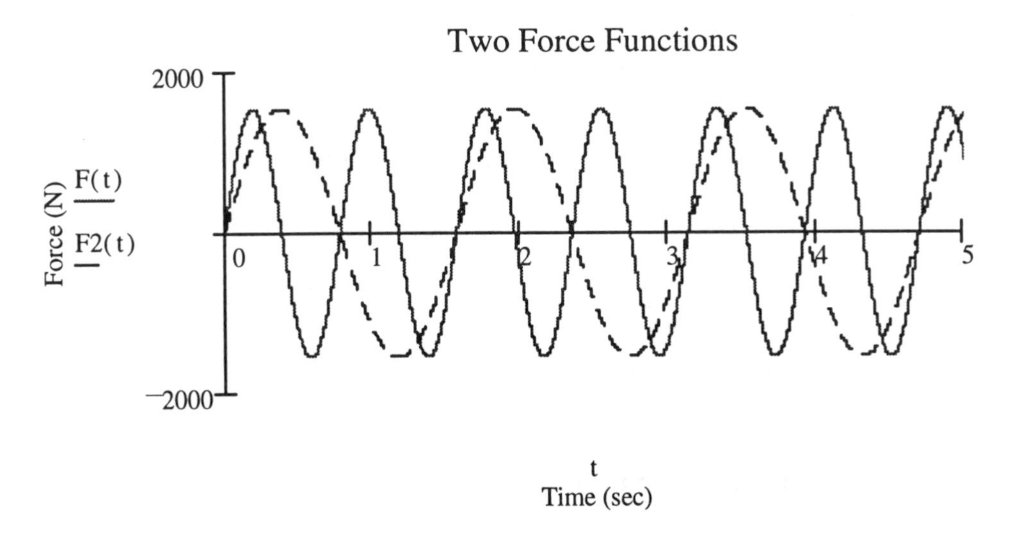

We can see that the principle difference between the two functions is their periods. $F(t) = 1{,}550 \sin 8t$ has a period of $\pi/4$, while $F_2(t) = 1{,}550 \sin 4t$ has a period of $\pi/2$. The first function, $F(t)$, completes two cycles in the time the second function, $F_2(t)$, completes only one. The amplitude of the two functions is the same (1550).

B. Using the new forcing function, $F_2(t)$, with all other conditions remaining the same, the equation of motion becomes:

$$850\, y''(t) + 13{,}600\, y(t) = 1550 \sin 4t$$

Solving this differential equation we note that now $w = w_0$. The homogeneous part of the solution remains the same:

$$y_c(t) = c_1 \cos 4t + c_2 \sin 4t$$

But, in finding the characteristic equation of the nonhomogeneous part, we are led to a solution ($y_p(t) = c_3 \cos 4t + c_4 \sin 4t$) which is absorbed by the homogeneous solution. Therefore, we multiply this expression by the variable t and obtain the form for the particular solution:

$$y_p(t) = c_3 t \cos 4t + c_4 t \sin 4t$$

We see:

$$y_p'(t) = -4 c_3 t \sin 4t + c_3 \cos 4t + 4 c_4 t \cos 4t + c_4 \sin 4t$$

$$= (c_3 + 4c_4 t) \cos 4t + (c_4 - 4c_3) \sin 4t$$

$$y_p''(t) = -16 c_3 t \cos 4t - 4 c_3 \sin 4t - 4 c_3 \sin 4t - 16 c_4 t \sin$$

$4t$

$$+ 4 c_4 \cos 4t + 4 c_4 \cos 4t$$

$$= 8 [(c_4 - 2c_3 t)(\cos 4t) - (c_3 + 2c_4 t)(\sin 4t)]$$

Now, substituting back into the original equation:

$$850[8 [(c_4 - 2c_3 t)(\cos 4t) - (c_3 + 2c_4 t)(\sin 4t)]] + 13{,}600[(c_3 t)(\cos 4t) + (c_4 t)(\sin 4t)] = 1550 \sin 4t$$

To solve for the coefficients we group into like terms:

$$(6800 c_4)\cos 4t + (-6800 c_3 - 1500)\sin 4t = 0$$

Substituting $t = 0$, we see that $c_4 = 0$. This makes sense since the forcing function has no cosine term and there are only derivatives of even order. The coefficient of the cosine term in this equation should be zero.

Letting t be any other value we see that $(-6800 c_3 - 1500)$ must always be zero. Therefore, $c_3 = -0.2205$.

The particular solution becomes: $y_p(t) = -0.2205 t \cos 4t$

Forming the general solution by combining the complementary and particular solutions, we obtain:

$$y(t) = c_1 \cos 4t + c_2 \sin 4t - 0.2205t \cos 4t$$

Now use the initial conditions, $y(0) = 0$ and $y'(0) = 0$, to solve for the coefficients:

$$y(t) = c_1 \cos 4t + c_2 \sin 4t - 0.2205t \cos 4t$$

$$y(0) = c_1 + 0 - 0 = 0$$

$$c_1 = 0$$

$$y'(t) = -4c_1 \sin 4t + 4c_2 \cos 4t + 0.882t \sin 4t - 0.2205 \cos 4t$$

$$y'(0) = 0 + 4c_2 + 0 - 0.2205 = 0$$

$$c_2 = 0.055125$$

Substituting into the general solution:

$$y(t) = 0.055125 \sin 4t - 0.2205t \cos 4t$$

C. The following graph is obtained by plotting the resultant motion equation above:

Graph of $y(t) = 0.055125 \sin 4t - 0.2205t \cos 4t$

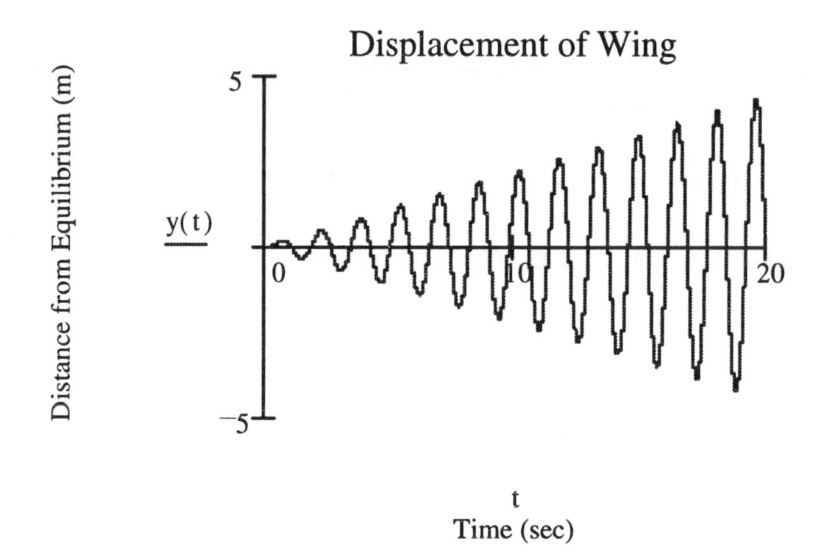

D. We can see by the graph that as t grows large, the displacement of the center-of-mass of the wing (the amplitude of the wing flutter) grows larger in each period (cycle). As described in the background material, the phenomena occurring is called mechanical resonance. This resonance will eventually result in the breaking off of the wing of the aircraft, unless other forces come into play at these larger amplitudes which change the governing differential equation of motion.

Part 3.

With the wing in free fall toward the earth, we can use Newton's Second Law to model the equation of motion:

$$\Sigma \text{ Forces} = ma = F_g + F_d$$

$$m\,y''(t) = mg - k_1 y'(t) - k_2 (y'(t))^2$$

A. In terms of modeling the velocity of the wing, $v(t) = y'(t)$. The differential equation becomes:

$$m\,v'(t) = mg - k_1 v(t) - k_2 v^2(t)$$

or

$$v'(t) + (k_1 v(t) + k_2 v^2(t))/m = g$$

Since we know $k_1 = 10.5$, $k_2 = 6.5$, $m = 850$ kg, and $g = 9.8$ m/s^2

The equation of motion becomes: $v'(t) + (10.5v(t) + 6.5v^2(t))/850 = 9.8$

With initial conditions: $v(0) = 0, \quad v'(0) = 0$

This is a first order, non-homogeneous, non-linear differential equation.

**Note that we are ignoring any velocity in the horizontal direction as well as any rotational motion, and simplifying the problem by considering only the vertical components in our analysis.

B. Using Euler's Method (our selected numerical solution technique), we obtain the difference equation:

$$v_{n+1} = v_n + \Delta t\,[9.8 - (10.5v_n + 6.5v_n^2)/850]$$

Choosing a step size of $\Delta t = 0.01$ seconds, we can use a spreadsheet to determine the numerical solution (see attached spreadsheet). Using the spreadsheet data, we find that the vertical velocity of the wing 4 seconds after it snaps is 28.1 meters per second.

C. To determine what the velocity of the wing is after it has fallen 50 meters, we need to determine the time at which it reaches the 50-meter distance. We can extend the spreadsheet above (see attached spreadsheet) using the estimated velocity at a certain time to find the distance at that time using the formula:

$$d_{n+1} = d_n + \Delta t\,(v_n)\,.$$

At 3.43 seconds the wing had fallen 50 meters. Therefore, we obtain the velocity at 3.43 seconds from the velocity spreadsheet. The velocity of the wing after falling 50 meters is 25.889 meters per second.

D. At terminal velocity the rate of change of velocity, $v'(t)$, becomes 0, so
$0 = (9.8)(850) - 10.5v(t) - 6.5v^2(t)$. We can solve this quadratic equation to find that

$$v_{term} = 10.5 \pm \sqrt{\frac{(10.5)^2 + 4(6.5)(9.8)(850)}{2(-6.5)}}$$

$$= \{-36.62, 35\}$$

Taking the positive solution (the positive vertical direction being downward), we find the terminal velocity of the falling wing to be 35 m/s.

Part 4.

A. Damping forces would be at work on the real plane's wing. We assumed forces which damp the motion are negligible when in fact they probably are not. If damping is included, the motion of the wing will remain bounded. Then the question becomes whether or not the amplitude of vibration of the damped wing (which may still be quite large) is greater than the amplitude at which the wing will break. Damping forces might be provided by the actual structure of the wing or engine housing. Additionally, the impact of air in terms of lift and drag could also have a damping effect.

We modeled the situation using a spring-mass system model. That model uses a simplifying assumption that the relationship between the spring force and the displacement is linear. When the displacement becomes large, as in this case, the linear force law behind this spring-mass system is no longer applicable, and the problem becomes nonlinear.

B. Since damping force seems critical, the engineer would want to redesign the engine housing to better damp the vibrations. In addition to investigating forces acting on the wing, the engineer may want to analyze other items such as the size and shape of the wing, the design of the attachment of the wing to the aircraft, and the materials which make up the wing. Even more critical than damping forces is the relationship between the natural frequency of the wing and the frequency of the external driving force. The engineer can anticipate this by designing the wing to have a natural frequency that is higher than the engine frequency (if possible), or by designing the wing as a composite structure where structural components possess natural frequencies different from each other.

C. Flex in the wing could be modeled by considering the wing to be a system of piecewise linear members, each with its own spring constant. For example, a 2-piece wing might look like:

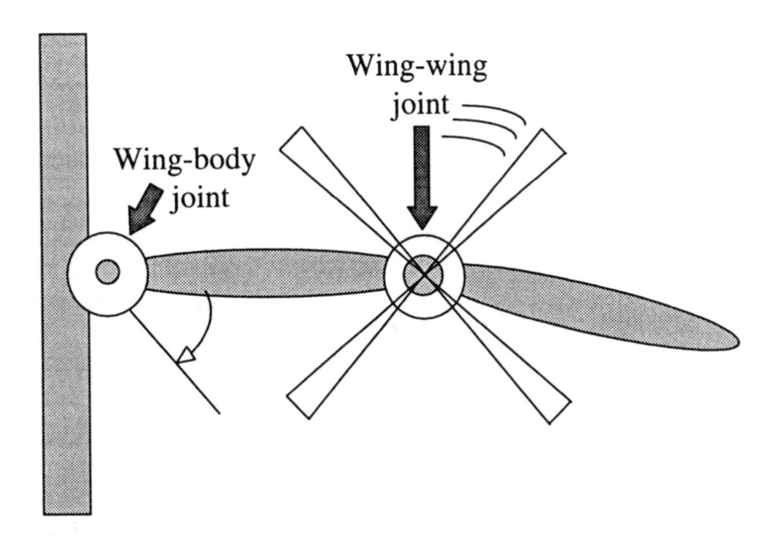

The equations are coupled through the requirement that the end of one section be at the same position as the beginning of the next section. This allows us to model the flex in the wing, with more sections allowing higher resolution of the spatial modes of vibration.

Title: Flying with Differential Equations

Notes for the Instructor

This project is designed to be used as the capstone activity for a differential equations block of instruction. It can be handed out at the end of the block and worked on intensively; in this mode, the project should take an average student (or student-team) about 8 hours to complete all requirements and to prepare a written report. However, we generally prefer to hand this out on the first day of the block, and to have a brief introduction and statement of the problem by a mechanical engineering or aeronautical engineering instructor (either in person when enrollment is small or on videotape when enrollment is large). The thrust of this introduction is that planes of the late 1950s had design problems, and that the students need to figure out why. The students have no clue how to do this yet (on the first lesson of the block), but by learning the differential equations material in the block they are beginning, they will have the tools they need to do the project. The engineer goes away, and we spend the block learning differential equations. As we cover skills that are used on the project, we discuss the connection and encourage the students to work on the project requirements in parallel, and to ask questions that arise along the way. On the second-to-last lesson of the block, the engineer returns, project teams brief their solutions and hand in their reports, and the engineer discusses extensions to the project and how learning differential equations will pay off for those students who have chosen or will choose engineering when they enter their program.

As discussed in the project background, this was a real problem that was pinpointed during a few famous crashes in the late 1950s; students find it fascinating that they can understand some simple ideas that engineers were wrestling with so recently, and that were a major concern of society at large.

This is a great project for challenging student intuition, and then demonstrating that intuition (though important) can be unreliable and not acceptable as a substitute for analysis. Survey the students at the beginning of the project about what they think the cause of the failures was. Most will answer (correctly), “engine frequency”. Follow up by asking what’s the path to failure; most will answer (incorrectly), that when the frequency gets high enough, the vibrations go out of control. Then ask how do they think engineers solved the problem; most will guess that they kept engines below a critical frequency, and that they made the wings stronger (both wrong). Revisit these questions during the project briefings; the role of the resonant frequency is now clear, and the process of designing a wing structure that incorporates components with several very different resonant frequencies makes sense as a design solution.

There are several threads present in the project that can optionally be picked up on and discussed or developed later as extensions. We generally discuss a few, but not all, of the following:

- Students sometimes think that spring-mass systems are contrived and of little use because they don't see them around much; the wing scenario demonstrates how a simple model can yield insight into a complicated situation, as it can when studying vibrations of atoms in a lattice, recoil of a cannon, response of an automobile suspension system, etc.

- You can discuss when it is acceptable to approximate rotational motion with linear motion (as we did here for small displacements of a long wing), and when we have to do it better by modeling the rotational motion (and how to do that).

- The open-ended question about modeling flex in the wing leads naturally into systems of Ordinary Differential Equations (ODEs). In the limit of a continuum of flex points, you could demonstrate how system of ODEs becomes a single Partial Differential Equation for the motion of the wing.

- We put a comment in the project about how modeling wind resistance with a v term, then a $v+v^2$ term, can be continued and that what we are really doing is constructing a Taylor Series approximation to the wind resistance. Spending a little time on this can clear up the empirical drag models they see (usually with no further explanation) in their physics courses.

Interdisciplinary Lively Application Project

Title: Planning A Backpacking Trip To Pikes Peak

Authors: Garrett Lambert
Beatrice Lambert
Michael Hendricks
Carol Harrison*
Gene Bennett*
Barry De Roos*
John Kellett*

Department of Mathematical Sciences and Department of Geography and Environmental Engineering, United States Military Academy, West Point, New York; Department of Mathematics, Susquehanna University, Selinsgrove, Pennsylvania; Department of Mathematics, University of Evansville, Evansville, Indiana; Department of Mathematical Sciences, Messiah College, Grantham, Pennsylvania

Authors with * after their name worked on this project at a week-long MAA Section workshop at Messiah College during the summer 1995

Editors: David C. Arney and Brian Winkel

Mathematics Classifications: Linear Regression, Data Analysis
Disciplinary Classifications: Exercise Physiology, Topography

Prerequisite Skills:

1. Solving Linear & Polynomial Regression Problems
2. Graphing
3. Scaling and Unit Conversion
4. Map Reading

Materials Available:

1. Problem Statement (6 Parts); Student
2. Sample Solution (6 Parts); Instructor
3. Topographic Maps; Student/Instructor
4. Notes for the Instructor

Computing Requirements: Polynomial Regression Tool

I. BACKGROUND

A hiker is planning a backpacking trip to the top of Pikes Peak in the Rockies outside Denver, Colorado. He has decided to hike along the Pikes Peak Toll Road. He has obtained a topographical map of the area showing the trail and each leg of the trip (see the bars drawn across the road trail on the map at Appendix A). The start point is labeled on the upper right part of the map sheet. The stop point, the top of Pikes Peak, is labeled at the bottom of the map. He has talked to a friend who is an exercise physiologist and found that he should exert a maximum of 3,000 kcal/day in hiking. His friend has also provided some experimental results from a study of the energy cost of human locomotion (see Tables 1 and 2 below). The hiker has determined that he will need to carry a pack weighing 40 pounds containing his food and shelter for the trip. Additionally, he must complete the trip in 2 days. The hiker weighs 180 pounds.

The hiker needs to answer the following questions in order to complete his planning for the trip:

1) Based upon the route indicated on the topographical map, how many kilocalories (kcal) will he expend in reaching the top of Pikes Peak? How long will it take him to complete the trip?

2) Assuming that he can expend a maximum of 3,000 kcal/day, where should he plan to make camp so that he does not exceed this limit? Can he make the trip in 2 days under this constraint?

3) Suppose that he decides to hike at most 7 hours per day. Where should he plan to make camp based upon this constraint? Can he make the trip within 2 days under this constraint?

4) If either of the two choices above is not possible, where are possible sites to camp and reach the mountain top within the constraints of kilocalories and hours hiked per day (i.e., relax the 2-day constraint)?

According to the exercise physiologist, the data in Tables 1 and 2 were gathered assuming that subjects moved at a comfortable rate of movement and carried a standard load of 40 pounds. The term "comfortable" is defined as the speed at which the subjects could move and not exceed a heart rate of 100 beats per minute or an energy expenditure of about 5 kcals/min.

Table 1: Table of kilocalories/kilometer by different weights of subject and angles of inclination

	0°	3°	6°	9°	12°	15°	18°	21°	24°	27°	30°
100 lbs	52.08	76.92	108.53	144.63	183.70	225.01	268.34	313.73	361.36	411.54	464.69
120 lbs	56.78	86.69	125.06	168.72	215.73	265.26	317.09	371.28	428.09	487.91	551.22
140 lbs	61.65	97.40	143.55	195.82	251.84	310.68	372.11	436.26	503.44	574.13	648.91
160 lbs	66.74	109.18	164.27	226.37	292.61	361.98	434.28	509.68	588.59	671.57	759.31
180 lbs	72.07	122.26	187.66	260.98	338.86	420.21	504.85	593.04	685.26	782.19	884.65
200 lbs	77.73	136.90	214.25	300.48	391.69	486.74	585.49	688.30	795.74	908.62	1027.91
220 lbs	83.77	153.48	244.78	345.95	452.56	563.41	678.44	798.09	923.08	1054.34	1193.02

Table 2: Table of kilometers per hour for a comfortable walk by different weights of subject and angles of inclination

	0°	3°	6°	9°	12°	15°	18°	21°	24°	27°	30°
100 lbs	5.50	3.73	2.64	1.98	1.56	1.27	1.07	0.91	0.79	0.70	0.62
120 lbs	5.05	3.31	2.29	1.70	1.33	1.08	0.90	0.77	0.67	0.59	0.52
140 lbs	4.65	2.94	2.00	1.46	1.14	0.92	0.77	0.66	0.57	0.50	0.44
160 lbs	4.30	2.63	1.75	1.27	0.98	0.79	0.66	0.56	0.49	0.43	0.38
180 lbs	3.98	2.34	1.53	1.10	0.85	0.68	0.57	0.48	0.42	0.37	0.32
200 lbs	3.69	2.09	1.34	0.95	0.73	0.59	0.49	0.42	0.36	0.32	0.28
220 lbs	3.42	1.87	1.17	0.83	0.63	0.51	0.42	0.36	0.31	0.27	0.24

Part 1: Analysis of Table 1--Calories per kilometer under different conditions

a) Plot a graph of the kcal/kilometer vs. angle of incline for several fixed weights (Suggest 120, 160, 200).

b) What are the similarities and differences between the graphs? Suggest reasons for these similarities and differences.

c) Examine the graphs and try to approximate the data with an elementary function.

Part 2: Analysis of Table 2--Kilometers per hour under different conditions.

a) Plot a graph of the kilometers/hr vs. inclination for several fixed weights (Suggest 120, 160, 200).

b) What are the similarities and differences between the graphs? Suggest reasons for these similarities and differences.

c) Examine the graphs and try to approximate the data with an elementary function.

Part 3: Analysis of the topographical map (see Appendix A):

a) Use the topographical map given (with specified route and trail segments) to construct a graph of the elevation vs. the distance traveled on the trail. Use straight lines to connect the points of the trail segments on the route.

b) Locate each of the changes in slope (at the trail segments) and describe what event occurs at that point.

Part 4: Construct a table to tabulate the results of the analysis of the graphs done in Requirements 1, 2, and 3. The table should have the following columns:

Table 3: Results of analysis of topographic map

Segment	Distance (Km)	Cumulative Distance	ΔElevation (Km)	ΔElev/ΔDist	Incline (Degrees)	Energy Expended (Kcal)	Time (Hours)

Part 5:

a) Based upon the route indicated on the topographical map, how many kilocalories (kcals) will he expend in reaching the top of Pikes Peak? How long will it take him to complete the trip?

b) Assuming that he can expend a maximum of 3,000 kcals/day, where should he plan to make camp so that he does not exceed this limit? Can he make the trip in 2 days under this constraint?

c) Suppose that he decides to hike at most 7 hours per day. Where should he plan to make camp based upon this constraint? Can he make the trip within 2 days under this constraint?

d) If either of the two choices above is not possible, where are possible sites to camp and reach the mountain top within the constraints of kilocalories and hours hiked per day (i.e., relax the 2-day constraint)?

Part 6:

What additional factors could be considered to make the model more realistic? How would these factors affect the results predicted above?

Appendix A: Topographic Map with trail segments labeled.

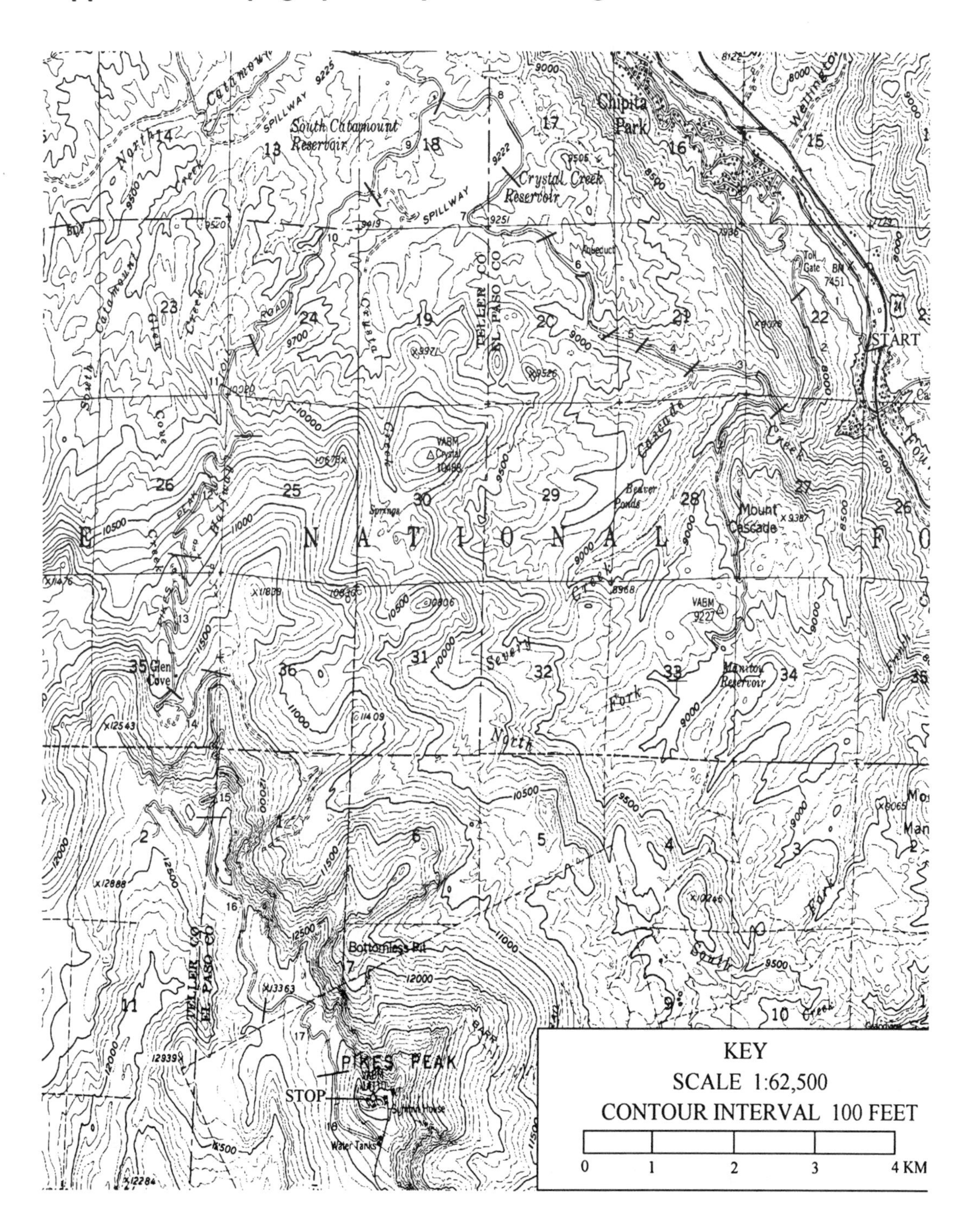

Title: Planning A Backpacking Trip To Pikes Peak

SAMPLE SOLUTION

Part 1: Analysis of Table 1--Calories per kilometer under different conditions

a) Plot a graph of the kcal/kilometer vs. angle of incline for several fixed weights (Suggest 120, 160, 200).

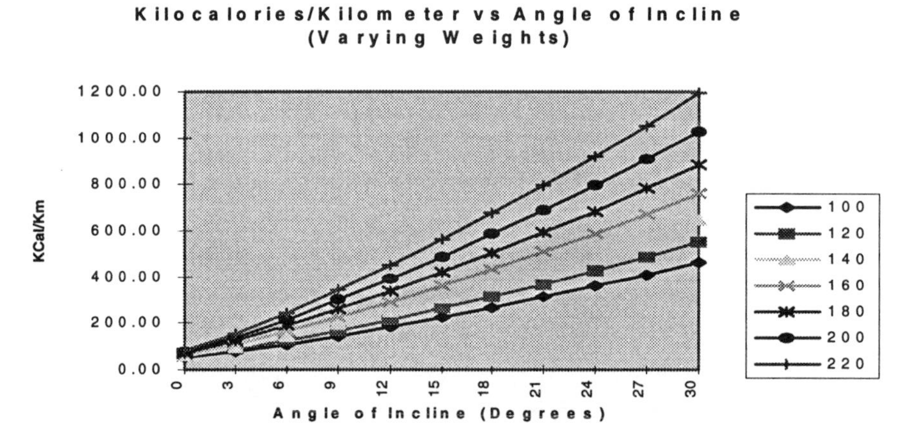

Figure 1.

b) What are the similarities and differences between the graphs? Suggest reasons for these similarities and differences.

The plots for each of the different weights indicate a similar trend--as we increase the angle of incline, the energy cost increases in a nonlinear fashion. Thus, within each weight class, there appears to be a non-linear relationship between the kcal/km expended and the angle of incline. Across the different weight classes, heavier subjects exhibit a higher energy expenditure due to the higher metabolic cost of moving a heavier body mass. It appears that the rucksack load of 40 pounds has a limited effect across the different weight classes. As a percent of body weight, we move from a high of 40% for a 100 pound person, to a low of about 18% for a 220-pound person for a 40-pound load. However, the energy cost is lowest for the lighter person.

c) Examine the graphs and try to approximate the data with an elementary function.

The shape of the plots indicates that we may be able to model the data adequately with a simple polynomial. A polynomial of order two or three may suffice to fit the data. We will fit three simple models to the data, briefly investigate their appropriateness, and select the one which best suits our needs. The proposed models are as follows:

Model 1: $y = \beta_0 + \beta_1 x + \varepsilon$

Model 2: $y = \beta_0 + \beta_1 x + \beta_2 x^2 + \varepsilon$

Model 3: $y = \beta_0 + \beta_1 x + \beta_2 x^2 + \beta_3 x^3 + \varepsilon$

where :

$y = \text{kcal / km expended}$
$\text{x} = \text{angle of incline}$
$\varepsilon = \text{random error}$
$\beta_i = \text{coefficients of independent variables}$

We assume that the ε are independently and identically distributed (i.i.d.), Normal (μ=0, constant variance=σ^2) random variables. Thus the expected value of ε for a given value of incline is zero. Since the hiker is 180 pounds, we will fit a model to the data for this weight. The parameters for each model were estimated using Microsoft Excel's Regression Tool which uses the method of least squares to estimate parameter values. A plot of the three models is shown in Figure 2.

It is apparent from the plot that all three models closely resemble the data. Of the three models, the first-order model seems to fit the data least well. The second and third order models do a better job. The difference between these two models is very small, and so we select the simpler model, the second-order model, as our function to approximate the data. The selected model is as follows:

Equation 1: $Energy = 65.548 + 19.638(incline) + 0.2571(incline)^2$

Equation 1 describes the energy (in kcal/km) required for a 180-pound person to negotiate an incline (measured in degrees) at a comfortable rate of movement. Similar analysis can be completed for each of the other weights.

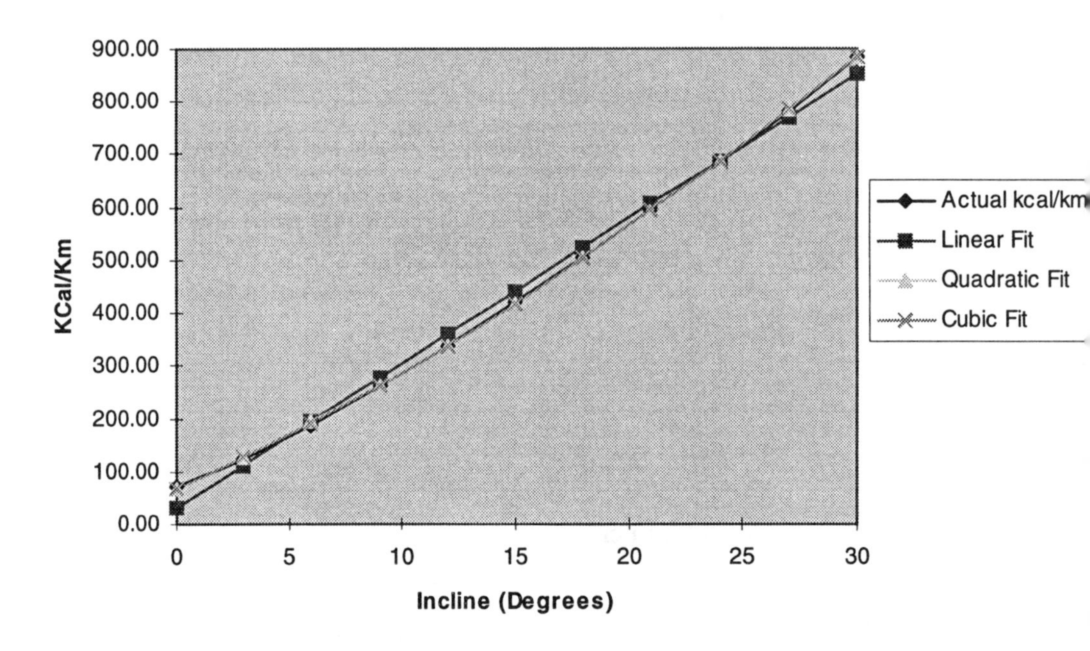

Figure 2.

Part 2: Analysis of Table 2--Kilometers per hour under different conditions.

a) Plot a graph of the kilometers/hr vs. inclination for several fixed weights (Suggest 120, 160, 200).

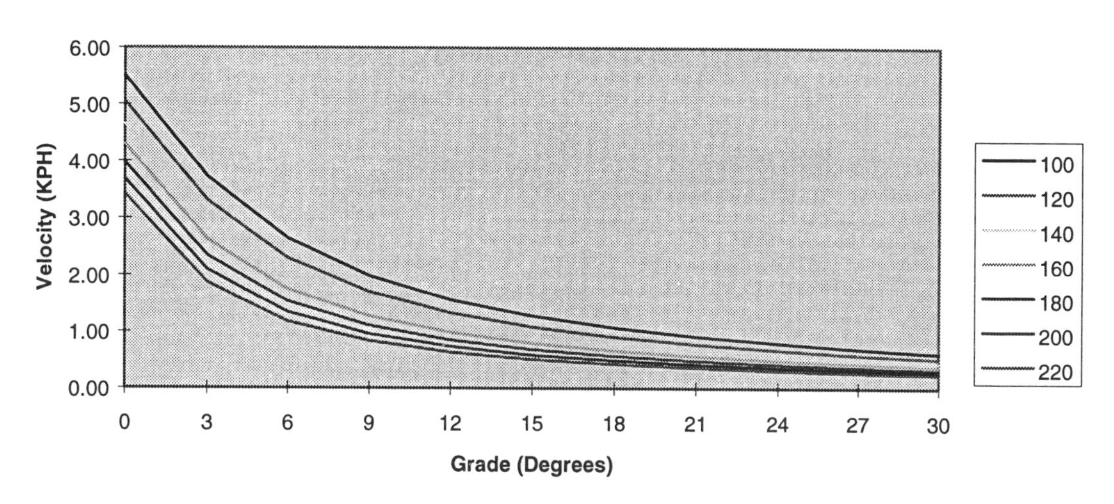

Figure 3

b) What are the similarities and differences between the graphs? Suggest reasons for these similarities and differences.

The plots suggest that, in order to maintain a comfortable level of energy expenditure as defined above, the hiker must decrease his rate of movement as the grade increases. It appears that a lighter person can maintain a higher rate of movement than a heavier person over the same incline.

c) Examine the graphs and try to approximate the data with an elementary function.

Using the same procedure as outlined in Requirement 1A, we fitted the following models to the data for a person weighing 180 pounds:

Model 1: $y = \beta_0 + \beta_1 x + \varepsilon$

Model 2: $y = \beta_0 + \beta_1 x + \beta_2 x^2 + \varepsilon$

Model 3: $y = \beta_0 + \beta_1 x + \beta_2 x^2 + \beta_3 x^3 + \varepsilon$

where :

y = velocity in kilometers per hour
x = angle of incline
ε = random error
β_i = coefficients of independent variables

A plot of the data and the fitted models is shown below.

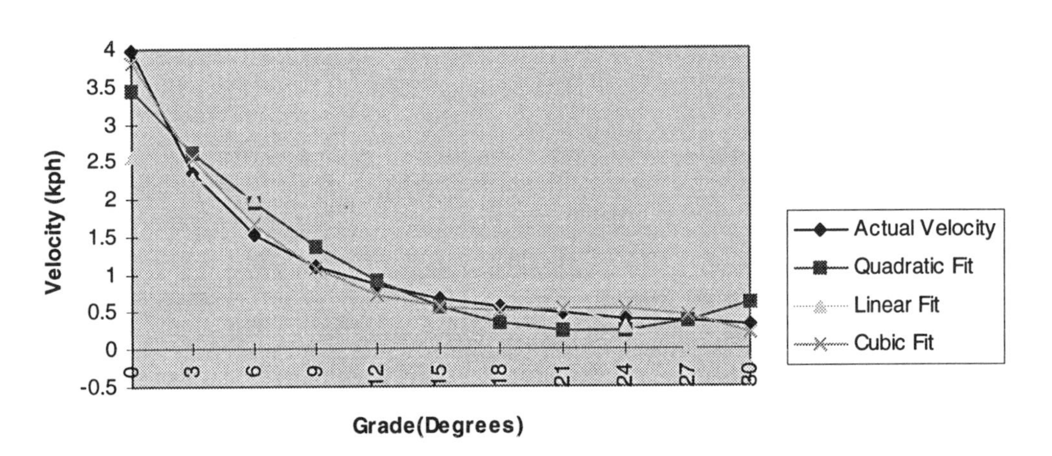

Figure 4.

Clearly Model 3, the cubic fit, best describes the data. Although a fourth-order model would do better, especially for the higher values of incline, the added benefit may not outweigh the increased complexity. Thus we will use the cubic model to predict the speed as a function of incline for a 180-pound hiker. The selected model is as follows:

Equation 2:

$$Velocity = 3.818 - 0.4927(incline) + 0.0244(incline)^2 - 0.0004(incline)^3$$

where *Velocity* is measured in kilometers per hour and *incline* is measured in degrees.

Part 3: Analysis of the topographical map (see Appendix A):

a) Use the topographical map given (with specified route and trail segments) to construct a graph of the elevation vs. the distance traveled on the trail. Use straight lines to connect the points.

Note: The distance for each segment has been provided in Table 3 below. This is helpful because of problems of scale encountered when copying/printing the maps.

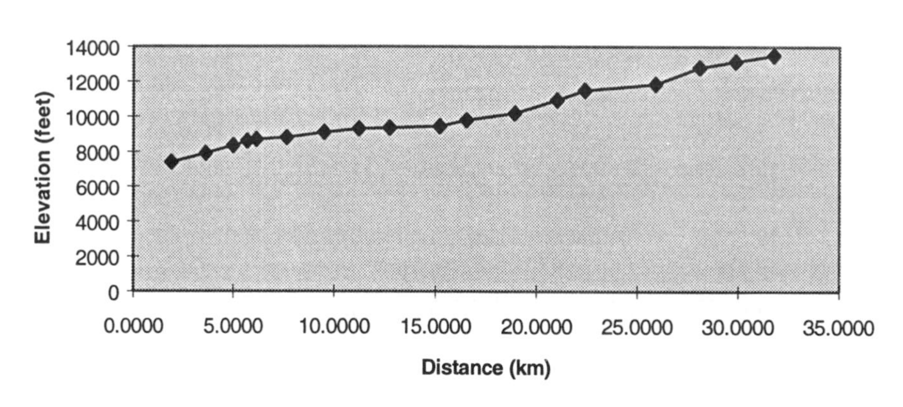

Figure 5.

b) Locate each of the changes in slope (at the trail segments) and describe what event occurs at that point.

Each change in slope represents a segment of the hiker's journey over which the incline is approximated. Thus each segment is simply the average rate of change of the incline over the distance traveled. As we reduce the size of each segment, we approach an instantaneous rate of change of the incline. This implies that our estimate of the metabolic cost of the hiker's trip will improve as we increase the number of segments we use to break up the journey.

Part 4: Construct a table to tabulate the results of the analysis of the graphs done in Requirements 1, 2, and 3. With proper conversions and calculations, the table should have the following columns:

Table 3: Results of analysis of topographic map(for a 180 pound hiker)

Segment	Distance (Km)	Cumulative Distance	ΔElevation (Km)	ΔElev/ΔDist	Incline (Degrees)	Energy Expended (Kcal)	Time (Hours)
1	1.9000	1.9000	0.1500	0.0789	4.5140	302.9187	0.9250
2	1.7000	3.6000	0.1350	0.0794	4.5404	272.0199	0.8308
3	1.4000	5.0000	0.0750	0.0536	3.0665	179.4580	0.5545
4	0.7000	5.7000	0.0300	0.0429	2.4540	80.7015	0.2546
5	0.4500	6.1500	0.0300	0.0667	3.8141	64.8844	0.1981
6	1.5000	7.6500	0.0900	0.0600	3.4336	204.0116	0.6257
7	1.8500	9.5000	0.0600	0.0324	1.8576	190.3908	0.6199
8	1.7000	11.2000	0.0150	0.0088	0.5055	128.4200	0.4755
9	1.5000	12.7000	0.0300	0.0200	1.1458	132.5784	0.4566
10	2.5000	15.2000	0.1050	0.0420	2.4050	285.6598	0.9030
11	1.3000	16.5000	0.1200	0.0923	5.2739	229.1457	0.7069
12	2.4000	18.9000	0.2250	0.0938	5.3558	427.4369	1.3207
13	2.1000	21.0000	0.1650	0.0786	4.4926	333.8180	1.0192
14	1.4000	22.4000	0.1050	0.0750	4.2892	216.3095	0.6598
15	3.5000	25.9000	0.2850	0.0814	4.6552	568.8823	1.7392
16	2.2000	28.1000	0.1050	0.0477	2.7325	266.4807	0.8316
17	1.7500	29.8500	0.1050	0.0600	3.4336	238.0135	0.7299
18	1.9000	31.7500	0.1830	0.0963	5.5015	344.5963	1.0681
TOTAL						4465.726	13.9191

Part 5:

a) Based upon the route indicated on the topographical map, how many kilocalories (kcal) will he expend in reaching the top of Pikes Peak? How long will it take him to complete the trip?

Table 3 shows that the hiker expended a total of about 4,466 kilocalories during the course of his trip to the top of Pikes Peak. An example of a calculation for the number of kcal expended (last column) for the first segment follows:

We input the angle of incline, 4.514°, into Equation 1 to obtain the number of kcal/kph expended : $Energy = 65.548 + 19.638(incline) + 0.2571(incline)^2$

$$Energy = 65.548 + 19.638(4.514) + 0.2571(4.514)^2$$

$$Energy = 159.4327 \quad \text{kcal / km}$$

We then multiply this result by the number of kilometers traveled for the first segment (1.9 km) to obtain a total of 302.92 kcal expended.

The total time required to complete the trip is about 14 hours. An example of a calculation of the time required to complete the first segment follows.

We input the angle of incline, 4.514°, into Equation 3 to obtain the speed of the hiker:

Equation 2:

$$Velocity = 3.818 - 0.4927(incline) + 0.0244(incline)^2 - 0.0004(incline)^3$$

$$Velocity = 3.818 - 0.4927(4.514) + 0.0244(4.514)^2 - 0.0004(4.514)^3$$

$$Velocity = 2.0543\,\text{kph}$$

Thus, traveling at a speed of 2.0543 kph for a distance of 1.9 km requires 0.925 hours.

b) Assuming that he can expend a maximum of 3,000 kcal/day, where should he plan to make camp so that he does not exceed this limit? Can he make the trip in 2 days under this constraint?

A table of cumulative number of kcal expended and time required for each segment will aid in answering this and the next question (see Table 4 below).

Table 4: Cumulative Energy and Time Expenditures (for a 180 pound hiker)

Segment	Distance (Km)	Cumulative Distance	Energy Expended (kcal)	Cumulative Energy Expended	Time (Hours)	Cumulative Time
1	1.9000	1.9000	302.9187	302.9187	0.9250	0.9250
2	1.7000	3.6000	272.0199	574.9386	0.8308	1.7559
3	1.4000	5.0000	179.4580	754.3966	0.5545	2.3103
4	0.7000	5.7000	80.7015	835.0981	0.2546	2.5649
5	0.4500	6.1500	64.8844	899.9826	0.1981	2.7630
6	1.5000	7.6500	204.0116	1103.9941	0.6257	3.3887
7	1.8500	9.5000	190.3908	1294.3849	0.6199	4.0086
8	1.7000	11.2000	128.4200	1422.8049	0.4755	4.4841
9	1.5000	12.7000	132.5784	1555.3833	0.4566	4.9407
10	2.5000	15.2000	285.6598	1841.0431	0.9030	5.8438
11	1.3000	16.5000	229.1457	2070.1888	0.7069	6.5506
12	2.4000	18.9000	427.4369	2497.6257	1.3207	7.8714
13	2.1000	21.0000	333.8180	2831.4437	1.0192	8.8906
14	1.4000	22.4000	216.3095	3047.7532	0.6598	9.5504
15	3.5000	25.9000	568.8823	3616.6355	1.7392	11.2896
16	2.2000	28.1000	266.4807	3883.1162	0.8316	12.1212
17	1.7500	29.8500	238.0135	4121.1297	0.7299	12.8512
18	1.9000	31.7500	344.5963	4465.7260	1.0681	13.9193

Table 4 indicates that the hiker should make camp at the end of the 13th segment in the vicinity of Glen Cove after having traveled 21 km and expending 2,831 kcal. This allows him to easily finish the last segment of 10.75 kilometers on the second day in about 5 hours with plenty of time to spare for sightseeing! Thus the hiker will be able to make the trip within two days under the 3,000 kcal/day constraint.

c) Suppose that he decides to hike at most 7 hours per day. Where should he plan to make camp based upon this constraint? Can he make the trip within 2 days under this constraint?

According to Table 4 above, if the hiker is restricted to stopping at the ends of trail segments, the hiker will have to make camp at the end of the

11th segment after a 6.6-hour hike during which he travels 16.5 km. This leaves him with 7.4 hours of travel time remaining at the end of the first day. Since this is greater than the 7-hour constraint, technically, the hiker cannot make the trip in two days under this constraint. Thus, on the second day he can go an additional 13.35 km in 6.3 hours and set up camp at the end of the 17th segment near the summit. On the third day he would walk the final 1.9 km in 1.1 hours to reach the top of Pikes Peak. Therefore, if the hiker is restricted to only 7 hours of travel per day, he will not quite reach the summit of Pikes Peak in two days.

d) If either of the two choices above is not possible, where are possible sites to camp and reach the mountain top within the constraints of kilocalories and hours hiked per day (i.e., relax the 2-day constraint)?

The time constraint of 7 hours per day is the more restrictive of the two constraints and therefore should dictate the location of the camp sites each night. The only adjustment needed to part 3 above is to plan for a third day of travel for the last segment consisting of 1.9 km in 1.1 hours.

Part 6

What additional factors could be considered to make the model more realistic? How would these factors affect the results predicted above?

Some additional factors to add realism:

- Obtain the functional relationship between energy expended and velocity so that the hiker's speed can be taken directly into account. The speed for Table 1 was assumed to be that which a person could maintain "comfortably" for the given incline and weight. Explicitly accounting for varying velocity adds a new level of realism to the problem.
- Break the trip into smaller segments so that the interval over which the grade is averaged is smaller and therefore more precise.
- Account for energy expenditure for movement downhill as well as uphill. Most hikes will include some downhill movement as well as uphill movement. Allowing a negative grade adds realism to the problem.
- There are a host of other factors that were not considered in our model (such as age, sex, and fitness levels of individuals) that will also influence the amount of energy required for the trip.

Title: Planning A Backpacking Trip To Pikes Peak

Notes for the Instructor

This modeling activity can offer students a very realistic application of data analysis and regression to determine a best-fitting polynomial and THEN using that fitted polynomial to help solve a backpacking problem by applying the polynomial function over a small interval and adding up the resulting energy costs, i.e. numerically integrating or accumulating the energy totals expended for various routes. This two-step approach is a common modeling activity: first get an empirical model (polynomial fit) and then use it.

The ideal way to use this material is to have students read the background first, preferably before class. Give some time in class for small groups (assuming the presence of classroom technology and expertise in using it for polynomial regression analysis) to assemble at least one of the polynomial models. Certainly have two groups do each of the polynomials of degree n = 1, 2, or 3. There can be reasoned discussion as to why the first-degree polynomial is ineffective, but there might be more discussion on the merits of the second- vs. third-degree polynomial. Traditionally we might take the lesser degree, but the computer spreadsheet would serve us well and with no additional cost to consider the third degree polynomial.

The teacher needs to get a final process, e.g. apply the energy expenditure model selects over EACH element of the path chosen, BEFORE class ends, and turn the mathematical analysis and accumulation effort over to the groups for construction of their report.

Students could obtain the necessary data on their own. Direct and indirect calorimetry methods would be required to obtain energy expenditures of test subjects similar to those given in Table 1 of the problem requirements. However, these methods of measuring energy expenditures can be involved and require equipment that is not usually available to undergraduate students. It turns out that there is a largely linear relationship between heart rate and energy expenditure. Therefore students could easily use heart rate to establish the "heaviness" of work and approximate the energy expenditure required using Table 5 below.

Table 5. Classification of Light to Heavy Work According to Energy Expenditure and Heart Rate

Classification	Total Energy Expenditure (kcal/min)	Heart Rate (beats/min)
Light Work	2.5	90 or less
Medium Work	5	100
Heavy Work	7.5	120
Very Heavy Work	10	140
Extremely Heavy Work	12.5	160 or more

Students could use Table 5 and a treadmill to obtain the data necessary to approximate the energy expenditure for different weights and angles of incline while carrying a load.

Another extension of this problem is to analyze Table 1 by plotting energy vs. weight for each of the different inclines. Based upon this analysis and the analysis conducted in Requirement 1, obtain the energy expended as a function of a person's weight and the angle of incline. Use this function to determine camp sites for a group of hikers of varying weights.

The breakdown of the trail (Pikes Peak Toll Road) into segments on the Map at Appendix A is arbitrary. The map provided in the problem statement is an example of one way to break the trail into segments. Students could be asked to decide for themselves the manner in which they will segment the trip. Clearly, as we increase the number of segments used (reduce the length of each leg), we improve the estimate of the energy and time requirements of the journey. Students can discover this for themselves if each student chooses a different number of segments and compares the results with other classmates. A topographic map without segments is provided at the end of this section in Appendix B. Of course other maps of local backpacking trails could be substituted for Pikes Peak.

Yet another extension to this problem is to determine the best route to take to the top of Pikes Peak among several competing alternatives. The best route could be defined as that route which requires the least amount of energy and/or time. There are several other trails on the map which students could use to analyze and decide which route is best for the given criteria.

Appendix B: Topographic Map of Pikes Peak (no trail segments)

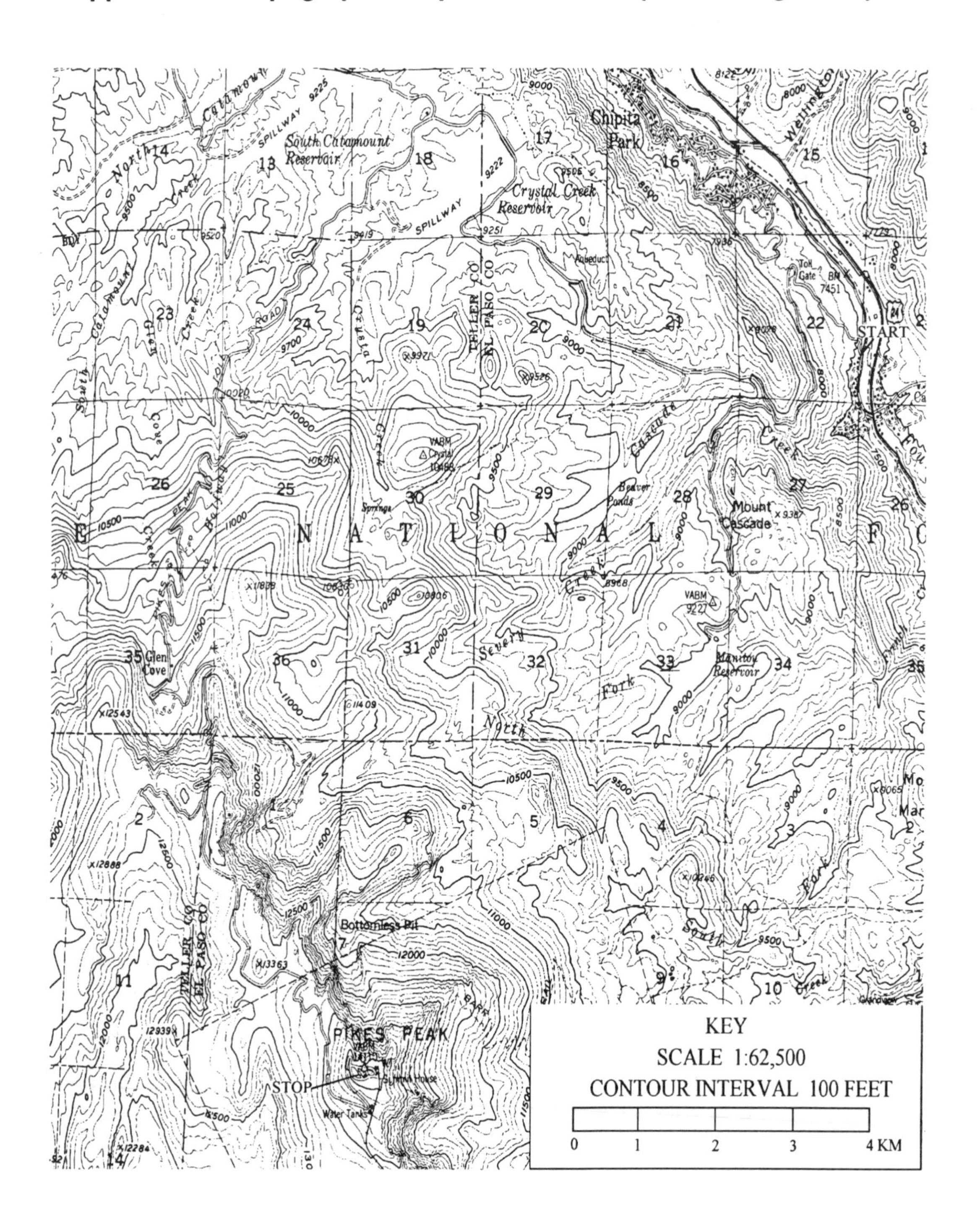

Interdisciplinary Lively Application Project

Title: SMOG in Los Angeles Basin

Authors: Charles Bass
Scott Torgerson
Jeffrey S. Strickland

Department of Mathematical Sciences and Department of Chemistry, United States Military Academy, West Point, New York

Editors: David C. Arney
Brian Winkel

Mathematics Classifications: Difference Equations, Matrix Algebra

Disciplinary Classifications: Chemistry

Prerequisite Skills:

1. Solving Systems of Difference Equations
2. Matrix Algebra
3. Solving for Eigenvalues and Eigenvectors

Physical Concepts Examined: Chemical Reactions

Materials Available:

1. Problem Statement (5 Parts); Student
2. Sample Solution (5 Parts); Instructor
3. Notes for the Instructor

Computer Requirements:

1. Computer Algebra System or Calculator (to find Eigenvalues and Eigenvectors)
2. Computation to Perform Iteration
3. Graphing Package

Photochemical SMOG permeates the Los Angeles basin most days of the year. While this problem is not unique to the Los Angeles area, conditions in the basin are well suited to this phenomenon. As early as 1542, explorer Juan Rodriguez Cabrillo named San Pedro Bay "the Bay of Smokes," because of the heavy haze from native fires that cover the area. The surrounding mountains and frequent inversion layers create the stagnant air that gives rise to these conditions. Automobiles feed the basin with the primary pollutants of hydrocarbons (RH) and nitrogen dioxide (NO_2). The ample sunshine drives atmospheric reactions that create the strong oxidants of ozone (O_3) and peroxyacetylnitrate (PAN). These oxidants are particularly destructive to human health, vegetation, and materials. Figure 1 shows the flow of pollutants into the LA Basin.

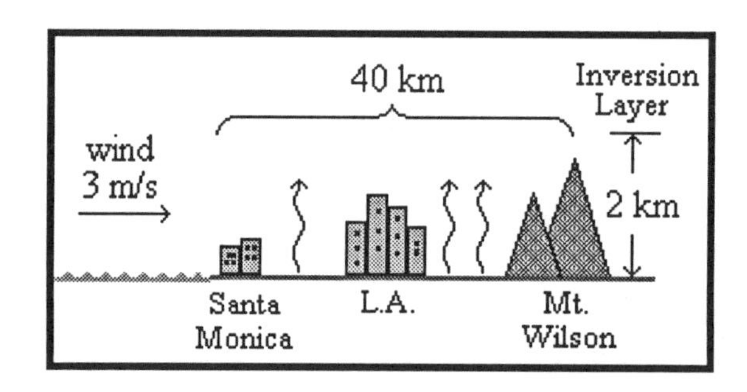

Figure 1: The development of SMOG through the entrant of pollutants.

Using a geometric "box model" approach, we can determine the steady state concentrations of the elements involved in the reaction between the pollutants and the sunshine. The "box model" considers the air volume defined by the area of the basin and the height of the inversion layer. Air enters at one end at the wind speed given, and exits at the other end. We assume that air in the "box" is perfectly mixed. As such, these conditions can be represented by the following "box model" shown in Figure 2.

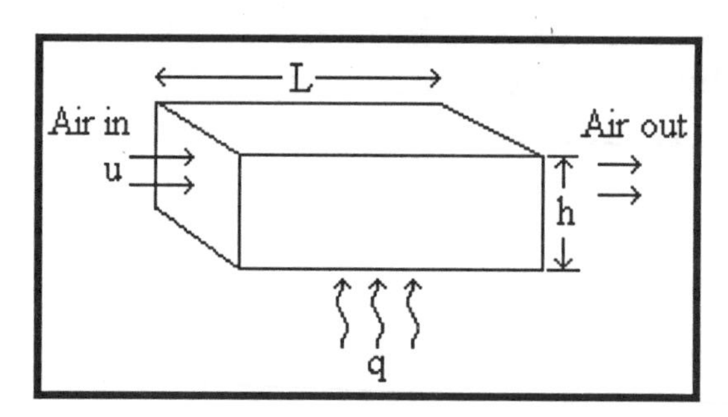

Figure 2: Box model for the production of SMOG in LA Basin.

In Figure 2 we let:

u = wind speed = 3 m/sec = 180 m/min
L = length of the box = 40,000 m
h = height of the box = 2,000 m
q = pollutants measured in moles/m^2 min

Consider the following conditions as they relate to the diagram in Figure 1 and the Box model in Figure 2. The amounts of pollutants (in moles) entering per square meter (m^2) per minute (min) for two pollutant chemicals are:

Hydrocarbons (RH) = 1.08×10^{-4} moles/m^2 min

Nitrogen Dioxide (NO_2) = 2.24×10^{-5} moles/m^2 min

Therefore, the steady state concentrations of these two pollutants can be determined by the formula: $C_{element} = \frac{qL}{uH}$, expressed in units of moles per liter of air. As such, the concentration of NO_2 is determined by:

$$C_{NO_2} = \frac{(2.24 \times 10^{-5}\ moles/m^2\ min)(40{,}000\ m)}{(180\ m/min)(2{,}000\ m)} \times \frac{1\ m^3}{100\ L} = 2.49 \times 10^{-9}\ \text{moles}/L$$

and the concentration of RH is:

$$C_{RH} = \frac{(1.08 \times 10^{-4}\ moles/m^2\ min)(40{,}000\ m)}{(180\ m/min)(2{,}000\ m)} \times \frac{1\ m^3}{100\ L} = 1.20 \times 10^{-8}\ \text{moles/L.}$$

The sun provides energy to the Los Angeles basin on a clear day at a maximum rate of $E_{max} = 5.0 \times 10^4\ J/m^2\ min$ (i.e., at noon). Since we recognize that the sun will not always radiate its maximum energy, even in Los Angeles, we generally assume an average energy radiated by $E_{ave} = 2.5 \times 10^4\ J/m^2\ min$. The energy from the sun (indicated by E) starts the following series of simplified reactions:

$$NO_2 \xrightarrow[k_1]{E} NO + O$$

$$O_2 + O \xrightarrow[k_2]{} O_3$$

$$O + RH \xrightarrow[k_3]{} R^O \text{ (a hydrocarbon radical)}$$

$$O_3 + RH \xrightarrow[k_4]{} R^O$$

$$NO_2 + R^O \xrightarrow[k_5]{} PAN$$

On the basis of these reactions and the original concentrations of the pollutants involved, the concentration of each element can be calculated at each time step (i.e. at discrete time intervals). To simplify the problem, we will consider the concentrations of NO_2, O_2, and RH to be constant because they are continuously being replenished in the atmosphere (C_{NO_2} and C_{RH} were determined and given previously). To simplify the notation, we will let $\boldsymbol{x(n) = C_O}$ at time n, $\boldsymbol{y(n) = C_{O_3}}$ at time n, $\boldsymbol{z(n) = C_{R^O}}$ at time n, and $\boldsymbol{w(n) = C_{PAN}}$ at time n. We also will let $d_1 = C_{NO_2} = d_2 = C_{RH}$, $d_3 = C_{O_2}$, and $d_4 = E_{AVE}$. We must identify each reaction step where an element appears. The rate of the reaction is the product of the concentration of all the reactants (left side) times the rate constant k_1. If the element appears on the right side, it is a product and is therefore created in the reaction. Finally, the concentration lost at the exit is given by u/L times the element. Given that the change in concentration of an element is equal to the concentration of the element CREATED due to reactions minus the concentration of the element CONSUMED due to reactions minus the concentration of the element LOST with exit from our "box model", we can determine the change in concentration for any of the remaining chemicals over any given time period. For example, the change of the concentration of oxygen (C_O) over the time period through one minute (time changing from *n* minutes to *n+1* minutes), is represented by the following equation, where C_O at time *n* minutes is represented by *x(n)*:

$$x(n+1) - x(n) = k_1 d_1 d_4 - k_2 d_3 x(n) - k_3 d_3 x(n) - \frac{u}{L} x(n).$$

This algebraically simplifies to

$$x(n+1) = \left(1 - k_2 d_3 - k_3 d_2 - \frac{u}{L}\right) x(n) + k k_1 d_4 d_1 .$$

We can also determine the concentration of the hydorcarbon radical (C_{RO}) at time *n* (given by *z(n)*), for example, by the same procedure producing the following equation:

$$z(n+1) - z(n) = k_3 d_2\ x(n) + k_4 d_2\ y(n) - k_5 d_1\ z(n) - \frac{u}{L} z(n)$$

which simplifies to

$$z(n+1) = k_3 d_2\ x(n) + k_4 d_2\ y(n) + \left(1 - k_5 d_1 - \frac{u}{L}\right) z(n) .$$

Since the oxidants involved, ozone (O_3) and peroxyacetylnitrate (*PAN*), are harmful to humans as well as the environment, it is important for the city to determine when these elements reach levels that will seriously affect those within the area. As the city consultant on this project, you are given the following assignments and questions:

Part 1:

a. Using the concentration of oxygen difference equation as an example, write the equations that will determine the concentrations of ozone (C_{O_3}) at time *n* (*y(n)*) and concentration of peroxyacetylnitrate (C_{PAN}) at time *n* (*w(n)*).

b. Write the system of four difference equations generated for *x(n), y(n), z(n),* and *w(n),* and express them using matrix notation.

Part 2: Given the following constants:

$d_1 = C_{NO_2} = 2.49 \times 10^{-9}$ *moles/L* $d_2 = C_{RH} = 1.20 \times 10^{-8}$ *moles/L.*

$d_3 = C_{O_2} = 0.0085$ *moles/L* $d_4 = E_{ave} = 2.5 \times 104\ J/m^2 min$

$k_1 = 2.0 \times 10^{-6} m^2/J$ $k_2 = 1.5$ *L/moles min*

$k_3 = 5.0 \times 10^4$ *L/moles min* $k_4 = 3.0 \times 10^5$ *L/moles min*

$k_5 = 1.0 \times 10^6$ *L/moles min*

and the assumption that the initial concentrations of (C_O, C_{O_3}, C_{RO}, and C_{PAN}) are zero at sunrise ($n = 0$).

a. Give a general solution to the system of dynamical equations. What happens to the concentrations of oxidants O_3 and *PAN* after a long period of time?

b. Make a graph of the concentrations of O_3 and *PAN* during a 12-hour day. (Remember *n* is minutes past sunrise.)

Part 3:

A high cloud layer attenuates half of the sunlight. How does this affect the long-term behavior of the oxidants? Be quantitative in your analysis.

Part 4:

A weather system reduces the height of the inversion layer to 1,000 meters. What is the impact on the long term behavior of the oxidants? Be quantitative in your analysis. (HINT: you will need to recalculate the concentrations of *RH* and NO_2.)

Part 5:

Comment on the validity of the model. Is the assumption that the concentrations of NO_2 and *RH* are constant reasonable? What is the impact on your model and your ability to solve it if these concentrations are not constant?

Title: SMOG in LA Basin

SAMPLE SOLUTION

Solution to Part 1. The system of equations that model the concentration of pollutants in the Los Angeles basis is

$$x(n+1) - x(n) = k_1 d_4 d_1 - k_2 d_3 x(n) - k_3 d_2 x(n) - \frac{u}{L} x(n)$$

$$y(n+1) - y(n) = k_2 d_3 x(n) - k_4 d_2 y(n) - \frac{u}{L} y(n)$$

$$z(n+1) - z(n) = k_3 d_2 x(n) + k_4 d_2 y(n) - k_5 d_1 z(n) - \frac{u}{L} z(n)$$

$$w(n+1) - w(n) = k_5 d_1 z(n) - \frac{u}{L} w(n)$$

Rearranging the equations gives us

$$x(n+1) = k_1 d_4 d_1 + x(n) - k_2 d_3 x(n) - k_3 d_2 x(n) - \frac{u}{L} x(n)$$

$$y(n+1) = k_2 d_3 x(n) + y(n) - k_4 d_2 y(n) - \frac{u}{L} y(n)$$

$$z(n+1) = k_3 d_2 x(n) + k_4 d_2 y(n) + z(n) - k_5 d_1 z(n) - \frac{u}{L} z(n)$$

$$w(n+1) = k_5 d_1 z(n) + w(n) - \frac{u}{L} w(n)$$

Simplifying these, we have

$$x(n+1) = \left(1 - k_2 d_3 - k_3 d_2 - \frac{u}{L}\right) x(n) + k_1 d_4 d_1$$

$$y(n+1) = k_2 d_3 x(n) + \left(1 - k_4 d_2 - \frac{u}{L}\right) y(n)$$

$$z(n+1) = k_3 d_2 x(n) + k_4 d_2 y(n) + \left(1 - k_5 d_1 - \frac{u}{L}\right) z(n)$$

$$w(n+1) = k_5 d_1 z(n) + \left(1 - - \frac{u}{L}\right) w(n)$$

Our matrix equation is

$$\begin{bmatrix} x(n+1) \\ y(n+1) \\ z(n+1) \\ w(n+1) \end{bmatrix} = \begin{bmatrix} \left(1-k_2d_3-k_3d_2-\frac{u}{L}\right) & 0 & 0 & 0 \\ k_2d_3 & \left(1-k_4d_2-\frac{u}{L}\right) & 0 & 0 \\ k_3d_2 & k_4d_2 & \left(1-k_5d_1-\frac{u}{L}\right) & 0 \\ 0 & 0 & k_5d_1 & \left(1-\frac{u}{L}\right) \end{bmatrix} \begin{bmatrix} x(n) \\ y(n) \\ z(n) \\ w(n) \end{bmatrix}$$

$$+\begin{bmatrix} k_1d_4d_1 \\ 0 \\ 0 \\ 0 \end{bmatrix}.$$

Solution to Part 2a.

We can use the coefficient matrix from Part 1 and substitute the given or determined constants.

This gives us the following coefficient matrix.

$$\begin{bmatrix} \frac{19643}{2000} & 0 & 0 & 0 \\ \frac{51}{4000} & \frac{9919}{10000} & 0 & 0 \\ \frac{3}{5000} & \frac{9}{2500} & \frac{99301}{100000} & 0 \\ 0 & 0 & \frac{249}{100000} & \frac{1991}{2000} \end{bmatrix}$$

or

$$R = \begin{bmatrix} 0.982150 & 0 & 0 & 0 \\ 0.01275 & 0.991900 & 0 & 0 \\ 0.0006 & 0.0036 & 0.993000 & 0 \\ 0 & 0 & 0.00249 & 0.9955 \end{bmatrix}.$$

We determine the following eigenvalues(using a calculator or computer package):

$$\left[\lambda_1 = \frac{19643}{20000} \quad \lambda_2 = \frac{9919}{10000} \quad \lambda_3 = \frac{99301}{100000} \quad \lambda_4 = \frac{1991}{2000}\right]$$

or

$$[\lambda_1 = 0.98215, \ \lambda_2 = 0.99191, \ \lambda_3 = 0.99301, \ \lambda_4 = 0.9955] \ .$$

Now we find an eigenvector associated with each of these eigenvectors. A CAS package gives the eigenvector associated with $\lambda_1 = \frac{19643}{20000}$ in the following form:

$$\left[x_1 = 1 \ \ x_2 = \ - \ \frac{17}{13} \ \ x_3 = \frac{890}{2353} \ \ x_4 = \ - \ \frac{166}{2353}\right]$$

or

$$[x_1 = 1 \ \ x_2 = \ -1.3077 \ \ x_3 = 0.3782 \ \ x_4 = -0.0705] \ .$$

For $\lambda_2 = \frac{9919}{10000}$, we have

$$\left[x_1 = 0 \ \ x_2 = 1 \ \ x_3 = -\frac{120}{37} \ \ x_4 = \frac{83}{37}\right]$$

or

$$[x_1 = 0 \ \ x_2 = 1 \ \ x_3 = -3.2432 \ \ x_4 = 2.2432] \ .$$

For $\lambda_3 = \frac{99301}{100000}$, we have

$$[x_1 = 0 \ \ x_2 = 0 \ \ x_3 = 1 \ \ x_4 = -1] \ .$$

For $\lambda_4 = \frac{1991}{2000}$, we have

$$[x_1 = 0 \ x_2 = 0 \ x_3 = 0 \ x_4 = 1] .$$

Therefore, the general solution of the homogeneous part of the system of difference equations is

$$c_1(0.9822)^k \begin{bmatrix} 1 \\ -1.3077 \\ 0.3782 \\ -0.0705 \end{bmatrix} + c_2(0.9930)^k \begin{bmatrix} 0 \\ 1 \\ -3.2432 \\ 2.2432 \end{bmatrix} + c_3(0.9949)^k \begin{bmatrix} 0 \\ 0 \\ 1 \\ -1 \end{bmatrix} + c_4(0.9955)^k \begin{bmatrix} 0 \\ 0 \\ 0 \\ 1 \end{bmatrix}$$

The solutions to the nonhomogeneous part of the system is found by solving the matrix equation

$$\mathbf{A} = \mathbf{RA} + \mathbf{B}, \text{ where } \mathbf{B} = \begin{bmatrix} 1.245 \times 10^{-10} \\ 0 \\ 0 \\ 0 \end{bmatrix}.$$

Using the inverse method for solving a linear system of equations, we have the computational formula

$$\mathbf{A} = (\mathbf{I} - \mathbf{R})^{-1}\mathbf{B} .$$

Computing this gives us the nonhomogeneous solution vector

$$A = \begin{bmatrix} 6.97478 \times 10^{-9} \\ 1.09788 \times 10^{-8} \\ 6.25303 \times 10^{-9} \\ 3.46001 \times 10^{-9} \end{bmatrix}.$$

Therefore, our general solution is

$$c_1(0.9822)^k\begin{bmatrix}1\\-\dfrac{17}{13}\\\dfrac{890}{2353}\\-\dfrac{166}{2353}\end{bmatrix}+c_2(0.9930)^k\begin{bmatrix}0\\1\\-\dfrac{120}{37}\\\dfrac{83}{37}\end{bmatrix}+c_3(0.9949)^k\begin{bmatrix}0\\0\\1\\-1\end{bmatrix}+c_4(0.9955)^k\begin{bmatrix}0\\0\\0\\1\end{bmatrix}$$

$$+\begin{bmatrix}6.9748\times10^{-9}\\1.0979\times10^{-8}\\6.2530\times10^{-9}\\3.4600\times10^{-9}\end{bmatrix}.$$

Now we want to use our general solution to determine what happens to the concentrations of oxidants O_3 (or $y(n)$) and *PAN* (or $w(n)$) after a long period of time. The matrix norm $\|R\|$ is 0.9955. Since $\|R\|$ is less than one, the equilibrium vector is stable, and the system will converge to this vector. Therefore, after a long period of time, the concentrations of oxidants O_3 (or $y(n)$) and *PAN* (or $w(n)$) will reach the equilibrium values 1.0979×10^{-8} and 3.4600×10^{-9}, respectively.

Solution to Part 2b.

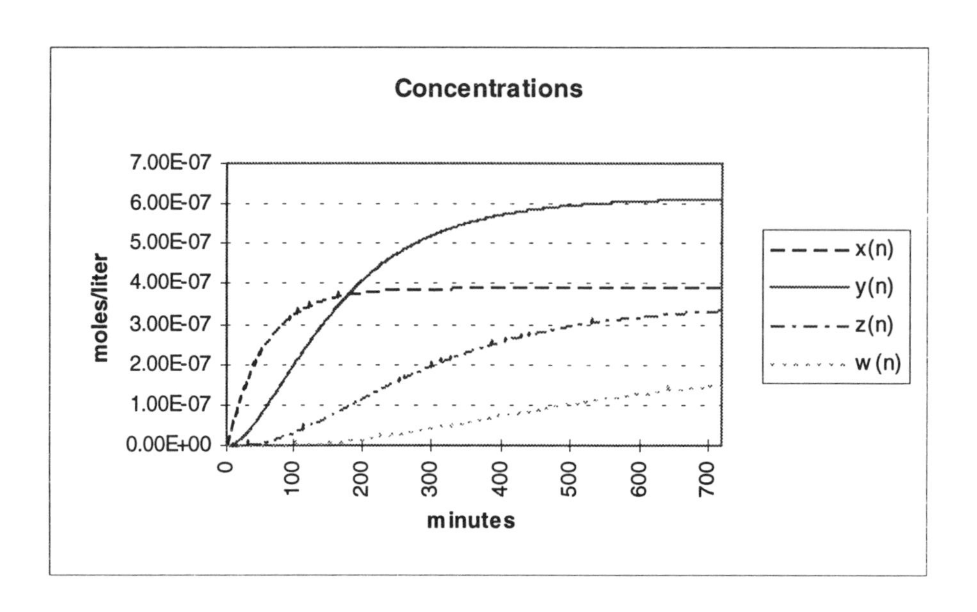

Solution to Part 3.

Reducing the energy of the sun by one-half affects the nonhomogeneous vector of our system. The original vector **B**, determined previously as

$$\begin{bmatrix} 1.245\times 10^{-10} \\ 0 \\ 0 \\ 0 \end{bmatrix} \text{ is changed to a new vector } \mathbf{B} = \begin{bmatrix} 6.225\times 10^{-11} \\ 0 \\ 0 \\ 0 \end{bmatrix}.$$

If we recompute the equilibrium vector **A**, we obtain the new nonhomogeneous part of the solution

$$\begin{bmatrix} 3.48739\times 10^{-9} \\ 5.48941\times 10^{-9} \\ 3.12651\times 10^{-9} \\ 1.73000\times 10^{-9} \end{bmatrix}.$$

Now we want to use our general solution to determine what happens to the concentration of oxidants O_3 (or $y(n)$) and *PAN* (or $w(n)$) after a long period of time. The matrix norm $\|R\|$ is 0.9955 has not changed. Since $\|R\|$ is less than one, the equilibrium vector is stable, and the system will converge to this vector. Therefore, after a long period of time, the concentrations of oxidants O_3 (or $y(n)$) and *PAN* (or $w(n)$) will reach new equilibrium values 5.4894×10^{-9} and 1.730×10^{-9}, respectively. The values of these concentrations at full sunlight were 1.0979×10^{-8} and 3.4600×10^{-9}, respectively. Therefore, the reduction in sunlight will cause the concentrations of oxidants O_3 (or $y(n)$) and *PAN* (or $w(n)$) to be reduced by one-half.

Solution to Part 4.

If a weather system reduces the height of the inversion layer to 1000 meters, then the concentration constants of NO_2 and *RH* (d_1 and d_2) are double their given values. The impact on the long-term behavior of the oxidants can be determined by resolving the system. When we do this, the matrix norm is still less than one, so the equilibrium vector is still stable. Therefore, we resolve for the equilibrium vector, which is

$$\begin{bmatrix} 3.37398 \times 10^{-9} \\ 3.67677 \times 10^{-9} \\ 3.21957 \times 10^{-9} \\ 3.56299 \times 10^{-9} \end{bmatrix}.$$

Therefore, after a long period of time, the concentrations of oxidants O_3 (or $y(n)$) and *PAN* (or $w(n)$) will reach 3.67677×10^{-9} and 3.56299×10^{-9}, respectively. The values of these concentrations, when the concentration constants of NO_2 and *RH* (d_1 and d_2) are not double their given values, were 1.0979×10^{-8} and 3.4600×10^{-9}, respectively. Therefore, the reduction in the height of the inversion layer to 1000 meters will cause the concentrations of oxidants O_3 (or $y(n)$) and *PAN* (or $w(n)$) to be reduced by 33.49 percent and increased by 102.98 percent, respectively.

Solution to Part 5.

The validity of this model is heavily dependent on the assumptions that the concentrations of NO_2 and *RH* (d_1 and d_2) are constant. However, this assumption is probably not reasonable. If the concentrations of NO_2 and *RH* (d_1 and d_2) are not constant, then our model would become nonlinear. For example, to determine the concentration of O_3 after n hours, we would have an equilibrium equation of the form

$$y(n+1) - k_2 p(n)x(n) + \left(1 - k_4 q(n) - \frac{u}{L}\right) y(n)$$

where $p(n) = C_{O_2}$, and $q(n) = C_{RH}$. We could iterate this new 6-equation, nonlinear system to analyze its short term behavior, but accurate analysis of the long-term behavior of this large, nonlinear system would be difficult.

Title: SMOG in LA Basin

Notes for the Instructor

This modeling activity can offer students a very realistic view of how large-scale modeling is done and how it is used to permit "what if" scenarios in policy analysis. The activity can serve a discrete dynamical or difference equation component of a course, or it may serve in a differential equation component of a course. Certainly if one does the discrete model, one should point out the modest jump to the continuous model, i.e. the differential equation, for students capable of making such a jump.

A concentration on the accounting, "CREATED, CONSUMED and LOST," is a major value in the modeling process and should be one of the major emphases in this project. This gives us something mathematical with which to work and should not be downplayed in any manner.

As to modeling of and application to the reality of the situation, the tweaking which goes on in Part 3 and the comparisons of resulting equilibrium values make the power of the model apparent. The teacher might wish to exaggerate the changes in some of the parameters to get drastic (albeit perhaps not too realistic) changes in the outputs. And certainly, combining effects such as reduction in sun (Part 3) AND altitude changes (Part 4) might offer further extensions.

Another use of the model is to report the various parameters, e.g. k_1, with a range of values and to ask for a range of outputs because such rate constants are often reported as ranges, and not as absolutes.

Part 5, on extending the model to a nonlinearity, is a worthwhile jumping off point to iteration as the way to obtain solution information. The teacher should be advised that iteration may be the appropriate behavior mode for solution at this point, coupled with plotting software to visualize the results of iterations.

A further analysis is to permit one of the rate constants or parameters to change with time, i.e. to introduce a non-autonomous term, e.g. the height of the box is slowly decreasing over time.

We presume students have access to the technology to obtain and easily manipulate the system and its attendant eigenvalues.

The ideal way to use this material is to have students read the background first, preferably before class. Give some time in class for small groups to assemble at least one of the difference equations from the assumptions, the teacher jumping in as necessary. Turn the problem of formulation over to the class, always calibrating between groups and encouraging sharing of information among groups, e.g., stop and ask a group to "present" their ideas. Try to get a final form of the model before class ends and turn the mathematical analysis over to the groups for construction of their report.

Also, convey that this model is just that -- a model. But emphasize that the model will permit local climatologists, policy makers, and economic developers to examine the issue of SMOG in terms of reasonable agreed-upon science and the mathematics to represent that science.

Interdisciplinary Lively Application Project

Title: Structural Mechanics - Beams and Bridges

Authors: John Scharf
Anthony Szpilka

Department of Mathematics, Engineering, Physics, & Computer Science, Carroll College, Helena, Montana

Editors: David C. Arney
Brian Winkel

Mathematics Classifications: Precalculus, Calculus

Disciplinary Classifications: Physics, Mechanics, Civil Engineering

Prerequisite Skills:

1. Modeling using Proportionality
2. Algebra of Linear Functions
3. Graphing and Symmetry
4. Data Analysis
5. Free Body Diagram and Moments

Physical Concepts Examined:

1. Static Equilibrium in 2 Dimensions
2. Conduct of an Experiment and Data Collection

Materials Available:

1. Problem Statement (12 Parts); Student
2. Sample Solution (12 Parts); Instructor
3. Notes for the Instructor

Computing Requirements: Spreadsheet and Graphing Tool are helpful, but not required

Material Requirements: Physics Laboratory with meter sticks, scales, and weights

Introduction:

There are numerous types of bridges. One type of bridge uses beams to transfer the weight of objects on the bridge to the supports. The supports for these bridges are usually columns or piers. Most highway bridges are of this type. On the other hand, suspension bridges use cables to transfer the weight of objects to the supports. The supports for suspension bridges are usually towers. Suspension bridges are used when the distance between the supports is very large. Familiar examples of suspension bridges are the Golden Gate Bridge in San Francisco and the Verrazano Narrows Bridge in New York.

You can obtain more information about bridges at the following address on the Internet: http://www.best.com/~solvers/bridge.html

Beams are structural members that support the weight of the objects resting on them by bending. This is in contrast to cables that support weights by stretching (see Figure 1). Considerable forces are required to bend a beam, whereas a cable offers little or no resistance to bending.

Compare, for example, a meter stick with a piece of rope that is one meter long. The meter stick is hard to bend, but the rope can easily be coiled into loops.

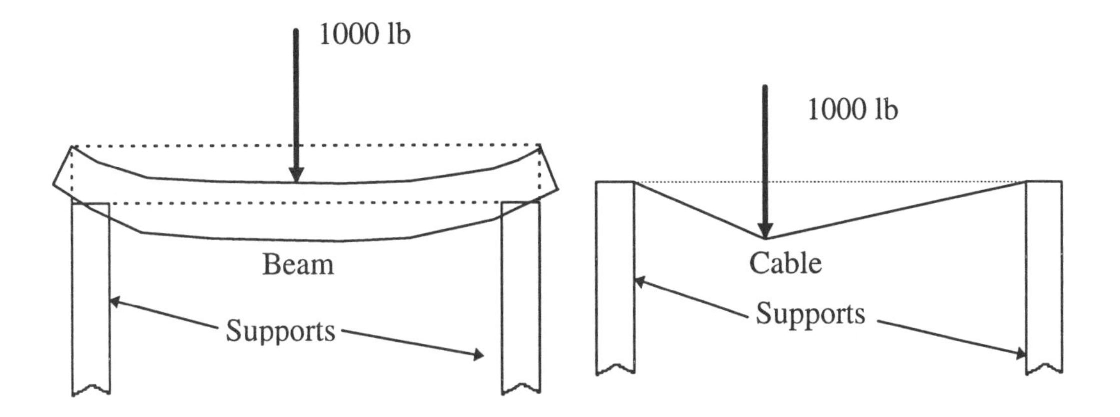

Figure 1 Bending of a Beam vs. Stretching of a Cable

In this project you will consider how the beams in beam-column bridges transfer the weight on the bridge to the supports. This is an important problem in the design of structures.

Single Span Bridges or Beams, with Simple Supports

A bridge beam that spans from one support to the other with no intermediate supports and no overhangs is said to be a single span beam (see Figure 2).

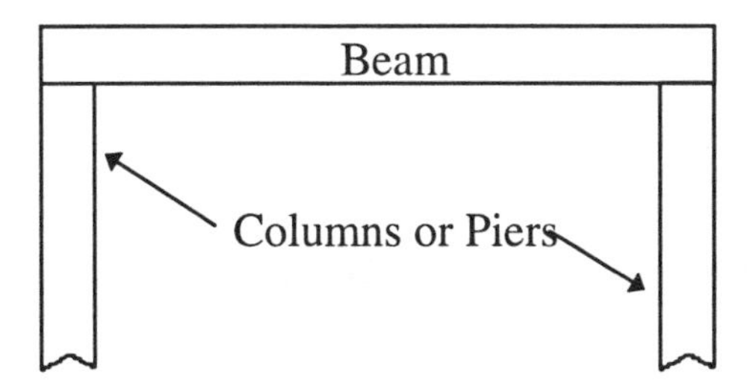

Figure 2 Single Span Beam

There are a variety of ways a beam can be connected to its supports. If the beam is simply resting on the supports, then we say that it is simply supported. In a real structure, the beam would most certainly have to be fastened to the supports so that it could not fall off on someone's head. The trick is to connect the beam and column together in such a way that the beam will push down (or pull up) on the column without bending it. If this is the type of connection, then the supports are simple supports. All of the beams in this project have simple supports.

For an example of a beam that was not adequately fastened to its supports, you may want to research the Kansas City Hyatt-Regency disaster. (see website http://ethics.tamu.edu/

What Is Your Conjecture?

It is important to realize that a significant portion of the total weight that a structure must support is the weight of the structure itself. For the time being, however, we will not include the weight of the structure. What we will do is put an object on the beam that weighs 1,000 pounds. In engineering this amount of weight is called a kilopound or a kip for short.

Part 1:

If the 1-kip object is placed at the center of the span, how much of the weight do you think will be distributed to each of the two supports? Why?

Part 2:

If the 1-kip object is placed directly over the left support, how much of the weight do you think will be distributed to each of the supports? Why?

Part 3:

If the 1-kip object is placed at the quarter point (see Figure 3), how much of the weight do you think will be distributed to each support? Why?

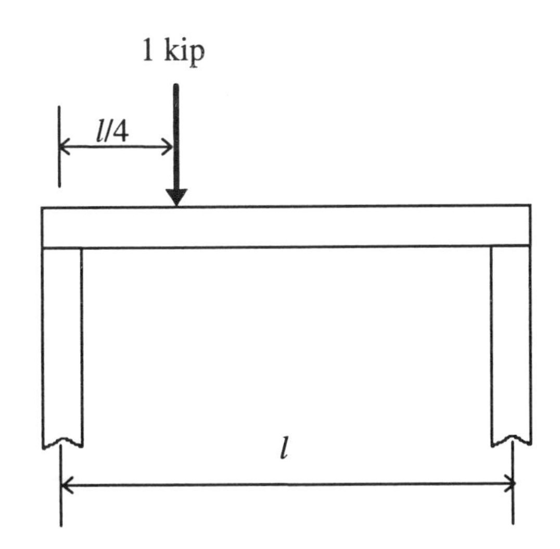

Figure 3 1 kip Weight at Quarter Point

Part 4:

How do you think the portion of the weight carried by the left support varies with the distance of the 1-kip weight from the left support? Write a mathematical expression which gives the weight carried by the left support as a function of the distance of the 1-kip weight from the left support.

Part 5:

How do you think the portion of the weight carried by the right support varies with the distance of the 1-kip weight from the left support? Write a mathematical expression which gives the weight carried by the right support as a function of the distance of the 1-kip weight from the left support.

Part 6:

If a weight is placed on a simply-supported, single-span beam, how does the beam distribute the weight to the supports? (Answer the question for the generic case shown in Figure 4.)

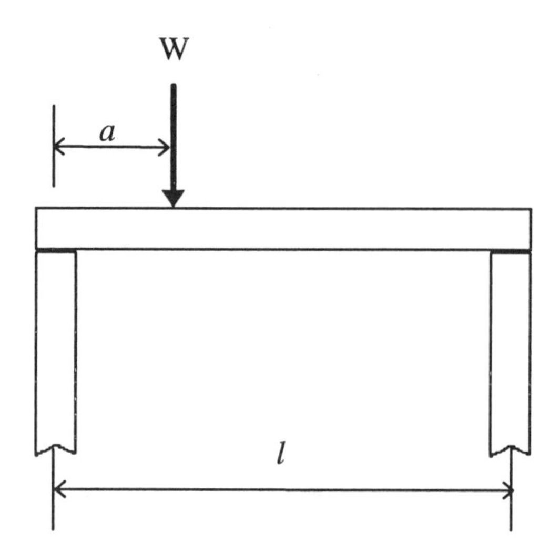

Figure 4 Generic, Simply-Supported, Single-Span Beam

Part 7:

Test Your Conjecture with an Experiment. In the physics laboratory there are meter sticks that you can use as beams, spring scales that measure height, weights, weight hangers, and supports. Devise an experiment to test your hypothesis about how the weight W is distributed to the simple supports of a single span beam. (See Part 8)

Part 8:

Prepare a report to describe the results of your experiment. The report should include a description of the experimental procedure, tables of recorded data, an analysis of the data, and a summary of your results and conclusions. Do the experimental results verify or contradict your a-priori conjectures about the support forces? If the experimental results contradict your conjectures, how would you modify your conjectures? Your report should be supported with figures, tables, and graphs.

Part 9:

Can you extend your theory to a simply-supported beam with an overhang as shown in Figure 5? In this case, a, the distance of the weight from the left support can be negative.

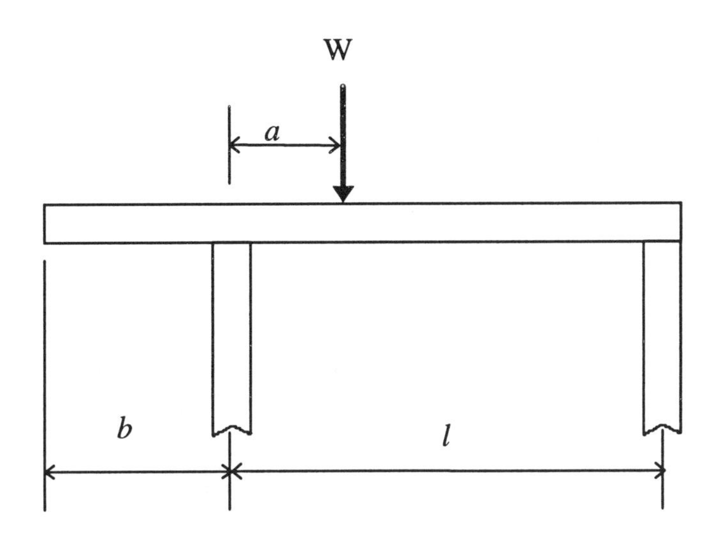

Figure 5 Simply-Supported Beam with an Overhang

Part 10:

Can you extend your theory to a beam with multiple loads as shown in Figure 6?

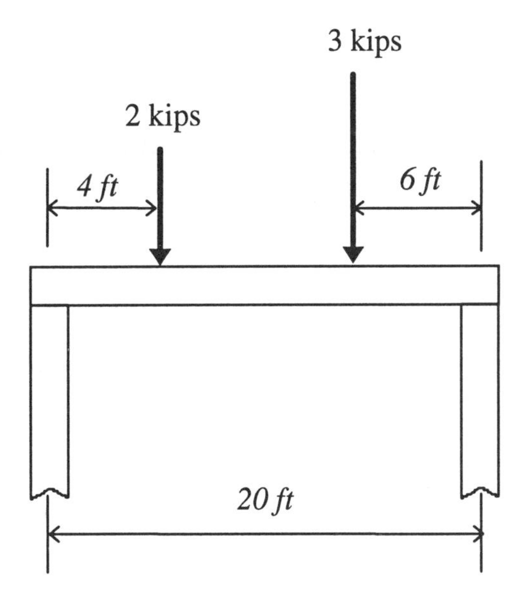

Figure 6 A Simply-Supported, Single-Span Beam with Multiple Loads

Part 11:

Use equilibrium principles from physics to determine the support forces for the case shown in Figure 4.

Part 12:

Figure 7 shows a beam with an intermediate support. Is it possible in this case to determine the support forces using equilibrium principles from physics? Why or why not?

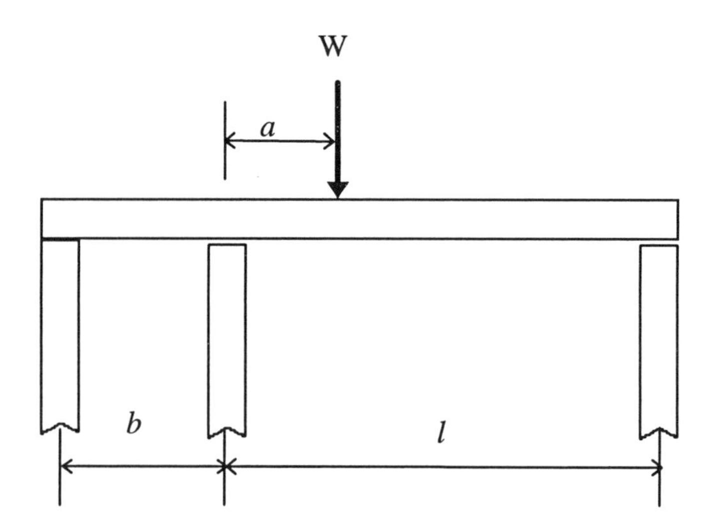

Figure 7 A Beam Spanning Continuously Over an Intermediate Support

Title: Structural Mechanics - Beams and Bridges

SAMPLE SOLUTION

Part 1:

Because of symmetry, it would seem reasonable that one-half or 0.5 kips of the 1-kip weight will be distributed to each support. An implicit assumption in this assertion is that the sum of the distributed weights is equal to the 1-kip weight.

Part 2:

Since the weight is directly over the left support, one might expect that all of the 1-kip weight will go to the left support and none will go to the right support.

Part 3:

In this case, more of the weight will be distributed to the left support than to the right because the weight is closer to the left support. One might guess that three-quarters of the weight or 0.75 kips will go to the left support and that one-quarter or 0.25 kips will go to the right support.

Part 4:

If the load were placed at the one-third point rather than the quarter point in the previous requirement, then the guess would be that two-thirds of the weight will go to the left support and one-third to the right. This seems to suggest that the weight that goes to the left support is proportional to the distance of the weight from the right support. If the distance from the left support is *a*, then the distance from the right support is $l-a$ and

$$W_{left} \propto (l-a) \text{ or } W_{left} = k(l-a) \ .$$

For the 1-kip weight, the constant of proportionality k can be determined from one of the specific cases in the previous requirements. For example, $W_{left} = 1\,kip$ when $a = 0$, which gives $k = \frac{1\,kip}{l}$. In summary, the portion of the 1-kip weight distributed to the left support is given by

$$W_{left} = 1\,kip\left(\frac{l-a}{l}\right) = 1\,kip\left(1 - \frac{a}{l}\right).$$

Part 5:

The symmetry of the structure suggests that the portion of the weight on the right support will be proportional to the distance of the 1-kip weight from the left support, which gives

$$W_{right} = 1\,kip\left(\frac{a}{l}\right).$$

Part 6:

If a weight of 2 kips is placed in the middle of the span, then half of the weight or 1 kip would be distributed to each support. If a weight of 4 kips is placed at the quarter point then one would expect that 3 kips would go to the near support and 1 kip to the far support. Extending this idea, if a weight of W is placed a distance a from the left support, then the weight would be distributed as follows:

$$W_{left} = W\left(\frac{l-a}{l}\right) \quad \text{and} \quad W_{right} = W\left(\frac{a}{l}\right).$$

Part 7:

Experiment Design: Use two spring scales to hang a meter stick from a support rack with the support points at 10 cm and 90 cm on the meter stick. The meter stick should have hanger clips at 10 cm increments along its length from 10 cm to 90 cm from which to hang a 200-gram (g) weight.

Before hanging the 200-gram mass from any of the clips on the meter stick, read the two spring scales and record the values. Hang the 200 g mass from each clip, in turn, and record the values on the two spring scales. To reduce the raw data, subtract the reading on the left spring scale, without the 200-g weight, from all of the other readings taken from the left spring scale, and do similarly for the right spring scale. Plot the reduced spring scale readings against the distance from the left support. Determine whether the results of the experiment verify or contradict your conjecture in Part 6.

Part 8:

Experimental Note: Even though a gram is not a unit of force (or weight), at a single location on the earth's surface, the weight of an object (here the 200-g weight) is proportional to its mass. In the equations above, the constant of proportionality can be canceled out allowing use of the mass as a measure of the weight.

The experimental procedure is described in Part 7. The left spring scale readings are denoted by SL and the right by SR. The readings on the spring scales without the 200-g weight hanging from the meter stick are denoted by SL0 and SR0. For each position of the weight, the measured portion of the 200 g weight that is distributed to the left and right supports is given by $WL = SL - SL0$ and $WR = SR - SR0$, respectively.

Recorded Data: $SL0 = 167$ g and $SR0 = 162$ g

Excel Spreadsheet with raw and reduced data:

	Raw Data		Distributed Weights	
a (cm)	SL (g)	SR (g)	WL (g)	WR (g)
0	367	162	200	0
10	342	188	175	26
20	316	214	149	52
30	290	236	123	74
40	266	262	99	100
50	242	285	75	123
60	216	310	49	148
70	192	336	25	174
80	166	362	-1	200

Excel Graph of WL and WR versus a:

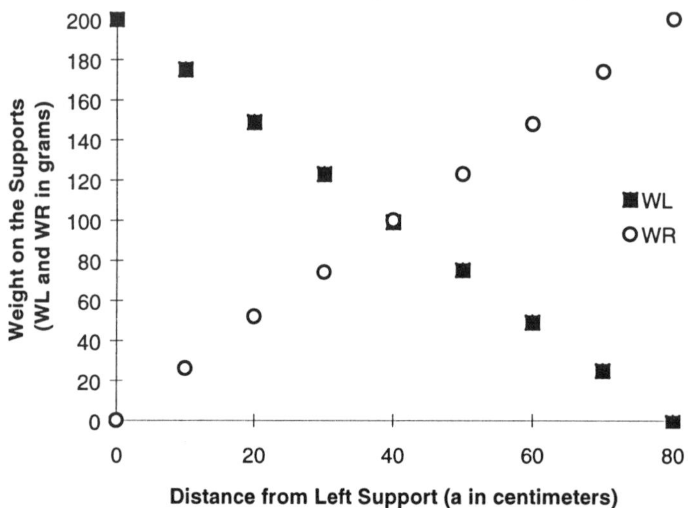

For comparison, the graph above showing the experimental values is re-plotted with the functions conjectured in Part 6, where $l = 80cm$ and $W = 200\,g$. The two functions in this specific case are:

$$W_{left} = 200g\left(1 - \frac{a}{80cm}\right) \text{ and } W_{right} = 200g\left(\frac{a}{80cm}\right).$$

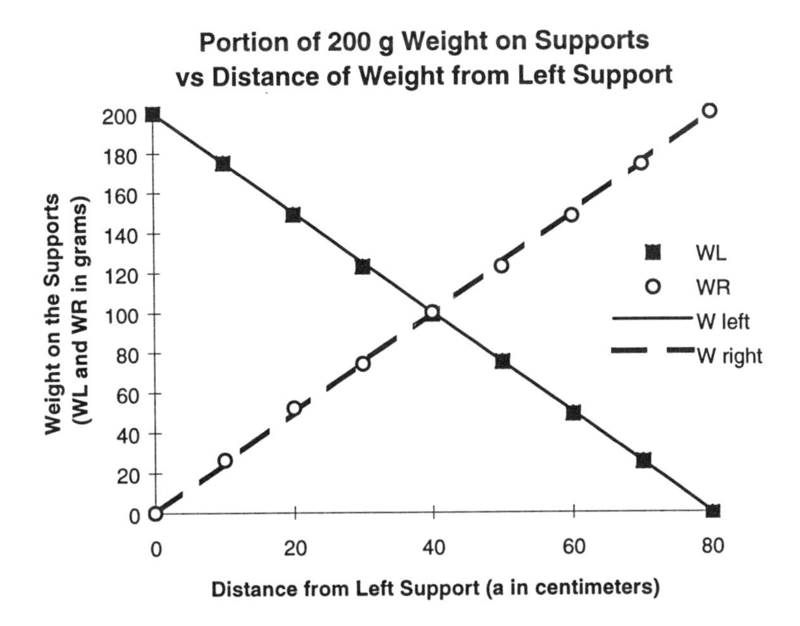

Because the experimental values are very close to the lines predicted by the conjecture, the experiment verifies the model.

Part 9:

To extend the model, one has to realize that the beam can pull up on a support as well as push down on it. (We assume that the supports are fastened to the beam so that it can, in fact, restrain the end of the beam from moving up.) In all of the cases considered in the preceding requirements, the weight on the beam causes the beam to push down on both supports. However, with the weight positioned on the overhang, it seems plausible that the beam could pull up on the right support. The model in Part 6 appears to accommodate this case if *a* is allowed to be negative. To illustrate, let $W = 200g$, $l = 80cm$, and $a = -10cm$ (i.e., the 200-g weight is placed on the overhang, 10 cm to the left of the left support). According to the conjecture, the portions of the 200-*g* weight that go to the supports would be:

$$W_{left} = 225g \text{ and } W_{right} = -25g .$$

The negative weight on the right support indicates that the effect of the 200-g weight on the overhang is to pull up on the right support with a force of 25 g. In this case the beam acts like a lever. Also note that the portions of the weight that go to the two supports still add up to 200 *g*.

Part 10:

To extend the theory to the case of multiple loads at various positions on the beam, the principle of superposition is used. This principle is based on the linearity of the two functions for W_{left} and W_{right}, and states that the weight distributed to each support can be calculated as the sum of the distributed weights from each object on the beam. For the beam in Figure 6, the weights distributed to the two supports would be:

$$W_{left} = 2\,kips\left(\frac{16}{20}\right) + 3\,kips\left(\frac{6}{20}\right) = 2.5\,kips$$

$$W_{right} = 2\,kips\left(\frac{4}{20}\right) + 3\,kips\left(\frac{14}{20}\right) = 2.5\,kips$$

Part 11:

Application of equilibrium principles requires a free body diagram of the beam. A free body diagram shows the beam in isolation from all other bodies with the other bodies replaced by the forces they exert on the beam. By Newton's Law of action and reaction, if the beam pushes down on a support, then the support pushes back up on the beam with an equal and opposite force. This force is the portion of the weight that the beam distributes to the support. The free body diagram for the beam in Figure 4 looks like this:

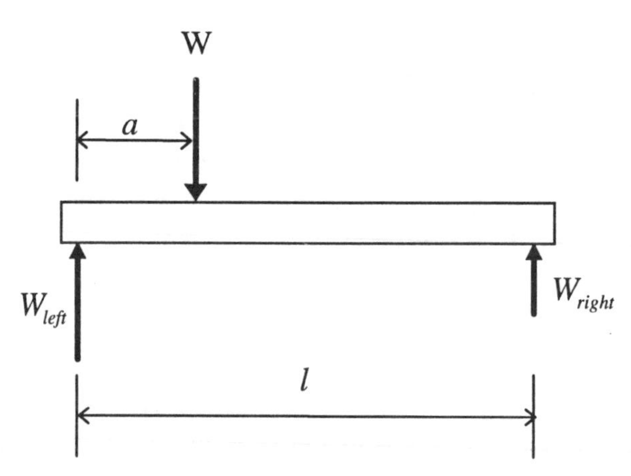

If the beam is in equilibrium, then the sum of the forces must be zero and the sum of the first moments of force about any point must also be zero. (The converse of this statement is also true.) The first condition gives the following equation: $W_{left} + W_{right} - W = 0$.

For the second condition, moments can be taken about the left support point to give:

$$\left(l \times W_{right}\right) - \left(a \times W\right) + \left(0 \times W_{left}\right) = 0$$

Solving these two equations for W_{left} and W_{right} gives:

$$W_{left} = W\left(1 - \frac{a}{l}\right) \quad \text{and} \quad W_{right} = W\left(\frac{a}{l}\right)$$

Part 12:

The free body diagram of the beam reveals the answer.

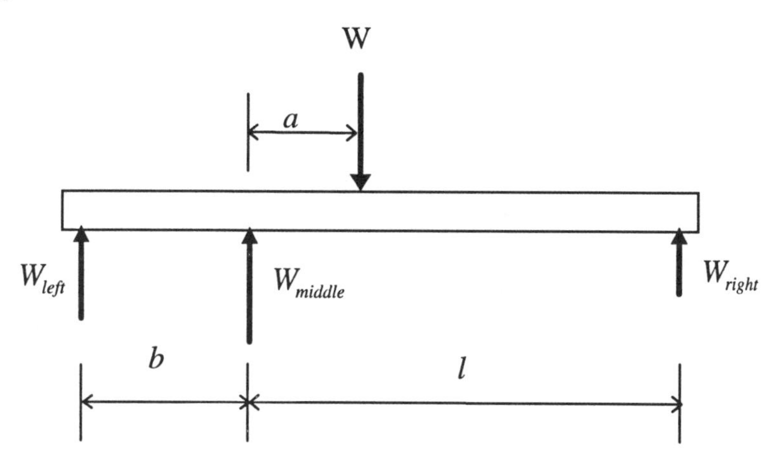

Now there are three unknown support forces, but there are only two equilibrium equations that apply in this situation. Consequently, there are infinitely many combinations of forces that satisfy the equilibrium equations. These are called self-balancing or self-equilibrating force systems. To determine the actual set of self-balancing forces that act on the beam in this situation, the deformations of the beam have to be considered. This gives an additional condition that allows determination of the correct support forces. Since the unknown forces cannot be determined with the principles of static equilibrium, this beam is said to be "statically indeterminate."

Title: Structural Mechanics - Beams and Bridges

Notes for the Instructor

"Structural Mechanics - Beams and Bridges" is the first project in a planned series of carry-through projects in structural mechanics. When completed, these subsequent ILAPs will be available from COMAP. This ILAP is appropriate for students in a precalculus, calculus, and/or an introductory physics course.

Two approaches can be used for this project. The first follows the sequence of Parts as they are written. In this first approach, the students are asked to make a conjecture about how a beam distributes a supported weight to its supports, and then to test their conjecture with an experiment. An alternative procedure would be to ask students to design and conduct an experiment to measure how a beam distributes a supported weight to the supports, and then to develop a mathematical model from the measured data. In the alternative approach, Parts 7 and 8 would be slightly modified and done first, followed by Parts 1 through 6 and then 9 through 12.

The prerequisite skills for this module include an understanding of proportionality, linear functions, and the ability to recognize geometric and algebraic symmetry. The last two Parts in this module require a knowledge of, or background research into, the conditions for static equilibrium of a two-dimensional rigid body. These conditions can be found in most introductory college-level physics texts or engineering statics texts.

Follow-through projects will include an extension to the cases of uniform and non-uniform distributed weights which would include the weight of the beam itself. This extension would require the use of integral calculus. The problem will be posed as a differential equation. The weight on the support can then be obtained by using the Green's function and the specified weight distribution. In structural mechanics, the Green's function is called an influence function. It measures the influence on the desired function (in this case, the weight that is distributed to the support) of a unit weight at a specified location on the beam.

Another follow-through project will consider the statically indeterminate case that is suggested in Part 12 of this first module. Given the displacement function for a simple beam, the weight distributed to redundant supports in an indeterminate structure can be determined. This is done by setting up a system of equations that applies forces to the beam at the support points to impose the support constraints, giving a system of simultaneous linear equations in the unknown support forces. These equations can be solved to determine the correct distribution of weights to the supports.

Interdisciplinary Lively Application Project

Title: Contaminant Transport

Authors: Richard Jardine
Michael Jaye
Chris King
Stan Thomas

Department of Mathematical Sciences and Department of Geography and Environmental Engineering, United States Military Academy, West Point, NY

Editors: David C. Arney and Brian J. Winkel

Mathematics Classification: Vector Calculus, Differential Equations, Partial Differential Equations
Disciplinary Classification: Environmental Engineering

Prerequisite Skills:

1. Modeling Space Curves with Vector Functions
2. Evaluating Line Integrals and Surface Integrals Using the Theorems of Green, Gauss, and Stokes
3. Modeling with the Diffusion Equation
4. Solving Partial Differential Equations (PDEs) Using the Method of Separation of Variables
5. Solving PDEs Using Numerical Methods
6. Performing Sensitivity Analysis and Interpreting Scenarios Using a Groundwater Flow Simulation Program.
7. Performing Sensitivity Analysis and Interpreting Scenarios Using a Contaminant Transport Simulation Program.

Materials Available:

1. Problem Statement (5 Parts, Parts 4-5 for Engineering Course); Student
2. Notes for the Instructor
3. Sample Solution (5 Parts); Instructor
4. Notes for the Instructor (Student Work on Group Projects)

Computing Requirements: SUTRA code for Parts 4 and 5

Part 1.

Prerequisite Skills:

1. Modeling space curves with vector functions
2. Evaluation of line and surface integrals using the theorems of Green, Gauss, and Stokes

Safe drinking water is a resource that many of us take for granted, but safe drinking water is a major concern of environmental engineers. In your first employment as an engineer in a county Public Works Office, you have been assigned to analyze groundwater problems related to an old waste disposal site.

A town within reasonable proximity of the waste disposal site had its water supply tested and learned that lead levels exceed the 15 micrograms per liter limit established by the Environmental Protection Agency. The town is considering litigation claiming the lead levels are due to contamination of the town's aquifer by wastes emanating from the waste disposal site.

The waste disposal had been used by federal and county agencies as a dumping site for most of this century, long before environmental awareness. It is well known that lead-cell vehicle batteries and lead-based paint cans were deposited at the dump prior to 1980, when toxic waste handling procedures were instituted on a county-wide basis as a result of the enforcement of the Resource Conservation Recovery Act of that year.

The County Engineer has tasked you to develop some analytical estimates preliminary to contacting private environmental engineering firms. The environmental section of the public works office is working on the problem, but is overwhelmed with other work.

Assignments:

A. A stream flows near the old waste disposal site toward the town and could be a contributing factor to the contamination. The change in elevation of the stream bed in the region of interest is from 375 m at point A near the landfill to 130 m at point B near the town's water treatment facility. Point B is 0.7 km west and 5.0 km north of Point A. One of the county engineers, for preliminary analysis and estimates, has modeled the stream bed using the position function $\vec{r}(t) = -0.7t\,\hat{\imath} + 5.0t^2\,\hat{\jmath} - 0.245t\,\hat{k}$ km.

(1) Discuss the appropriateness of the above vector function as a mathematical model of the stream bed. Sketch the model, preferably using a computer graphics package.

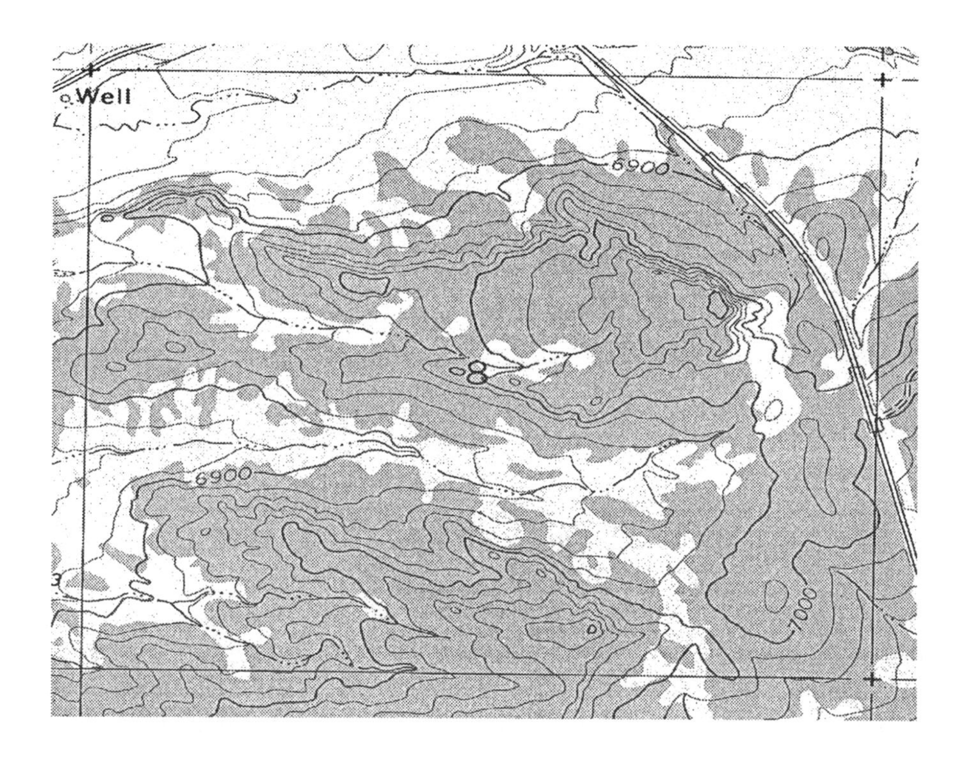

(2) Given that the water's velocity field is $\vec{V} = 0.01x\,\hat{\imath} - 0.05z\,\hat{\jmath} - 0.02y\,\hat{k}$ m/s and the density $\boldsymbol{r}$ of water is 1000 kg/m^3, find the flow of water moving in the stream. Interpret your result, including in the discussion the magnitude and the sign of the result, at a minimum.

B. Several bore holes have been dug near the waste disposal site, and instruments were placed in the wells to obtain data concerning the contamination of the aquifer by the landfill. In a 2-m long cylindrical portion of one of the holes, the groundwater velocity field is determined to be $\vec{V} = x\hat{\imath} + y\hat{\jmath} + 2z\hat{k}$ m/s.

(1) Determine the flux through the 2-m long portion of the well. The diameter of the well is 20 cm.

(2) Are there sources or sinks in that portion of the bore hole? Explain not only your answer, but also possible explanations for the result.

C. In another of the test wells, groundwater rotates around the center of the hole (with diameter 20 cm) with velocity $\vec{V} = \omega(y\hat{i} - 2x\hat{j})$, where *w*, the angular velocity of rotation, is 0.25 rad/s. If the circulation in the well exceeds $15\frac{kg-rad}{s}$, the data collected by instruments in the well will be corrupted. Determine whether or not the empirical results obtained from that hole are valid and justify your analysis.

Part 2.

Prerequisite Skills:

3. Modeling groundwater transport with the diffusion equation
4. Solving partial differential equations using the method of separation of variables

General Information: The movement of contaminants in groundwater is mainly affected by the processes of *advection* and *dispersion.* Advection describes the movement (mass transport) of contaminants due simply to the flow of water in which the mass is dissolved. The direction and rate coincides with that of the groundwater in which the contaminant is dissolved. Dispersion is the fluid process that causes a zone of mixing between adjacent fluids, occurring as a result of *diffusion* and mechanical dispersion. Diffusion is caused by random molecular motion in the solute driven by the spatial concentration gradient in the solute. For hydrogeological systems with fluid motion assuming small groundwater flow and therefore minimal mechanical mixing, the partial differential equation (PDE) used to model mass transport in one space dimension is the **advection-diffusion** equation,

$$D\frac{\partial^2 C}{\partial x^2} - v\frac{\partial C}{\partial x} = \frac{\partial C}{\partial t},$$

where $\boldsymbol{D}$ is the diffusion coefficient, $\boldsymbol{v}$ is the linear groundwater velocity, and $\boldsymbol{C}(x,t)$ is the concentration of the contaminant in mass per unit volume (micrograms/liter) at a position $\boldsymbol{x}$ at a time $\boldsymbol{t}$.

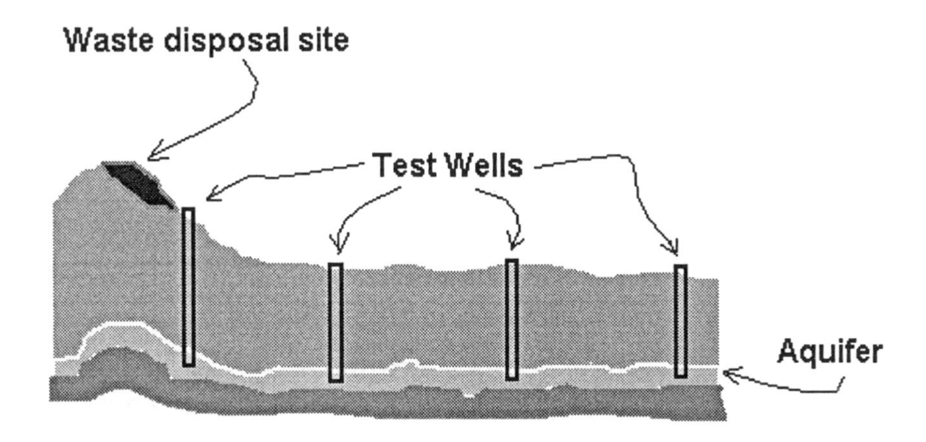

Assignments:

A. For the purposes of an initial estimate, the velocity of the groundwater is assumed negligible. The diffusion coefficient of the aquifer is 0.0008 km^2/day. The geological structure is as shown in the diagram. The landfill lies on the top slope of an elevated terrain feature. A Public Works Office old-timer has assured you that no lead left the waste disposal site after 1980. Recent EPA studies show that the lead levels at 10 km and beyond from the source are negligible due to dilution.

> (1) Write the PDE that models the given situation. Include in the model the initial and boundary conditions. Describe the meaning of each term in the equations in the context of the physical problem.
>
> (2) Discuss the assumptions made in using the model developed in (1). Describe the limitations the assumptions place on the model. Be sure to include justification of the boundary conditions you have selected for the model.

B. A harmless trace material has been injected into test wells in a manner which can be modeled by the initial condition (concentration of contaminant measured in micrograms/liter),

$$C(x,0)=\begin{cases}0, & 0\le x<1\\ x-1, & 1\le x<4\\ 6-\dfrac{3x}{4}, & 4\le x<6\\ \dfrac{15}{4}-\dfrac{3}{8}x, & 6\le x\le 10.\end{cases}$$

(1) Find and plot a Fourier series approximation to the initial condition. Graphically compare the approximation to the actual initial condition.

(2) Obtain a Fourier series solution to the boundary value problem subject to the given initial condition.

(3) Plot the solution obtained in (2) at $t = 0.0$ using the first 15 terms of the series. Graphically compare the result with the initial condition. Explain any differences between the plots of the two functions.

(4) Plot the first ten terms of the solution at $t = 0.0$, 5, 10, 25, and 50 years. Explain the physical phenomena described by the plots.

(5) Using 15 terms of the series solution, estimate the amount of trace material expected at the water treatment plant 5 km directly down gradient from the waste disposal site one year after the injection. How can this information be used to come to a determination of the effect of the waste disposal site on the aquifer?

(6) The initial condition ***C*(*x*,0)**, based on the insertion of trace material, is obviously a linear approximation. Discuss whether or not the approximation is physically relevant. Develop a more realistic model of the initial condition and discuss how that model affects the solution technique used above.

(7) Discuss how the modeling performed in parts (1) - (5) relates to the original problem of determining the contribution of the waste disposal site to the high lead levels in the town water supply, if it relates at all. What information would be needed to use this model? Old bore hole data is available in records maintained at the Public Works Office since 1980.

Part 3.

Prerequisite Skills:

5. Solving partial differential equations using numerical methods

Assignments:

A computer program is desired to obtain a numerical approximation for the concentration of lead in the groundwater using the full advection-diffusion equation for use in initial estimates.

A. Begin by comparing the numerical implementation with an accepted result.

(1) Develop a numerical algorithm that uses forward differencing in time (t) and centered differences in distance (x) for the PDE of part IIb.

(2) Implement the algorithm to solve the problem in part IIb using the following step sizes. Compare results:

(a) $\Delta x = 1$ km, $\Delta t = 1{,}000$ days;

(b) $\Delta x = 1$ km, $\Delta t = 500$ days.

(3) Estimate the trace material levels in the aquifer at $x = 5$ km and $t = 1$ year using the numerical algorithm. Graphically compare the result with the analytical solution obtained in part 2b, explaining any differences and how the differences might be decreased.

B. Develop a numerical implementation that includes advection.

(1) Develop an algorithm that numerically approximates the full advection-diffusion equation.

(2) Implement the algorithm and estimate the trace material using the numerical implementation. Graphically compare the results with the numerical solution obtained for the case without advection, explaining the effects of the advection term. Discuss the effect of changing $\Delta \mathbf{x}$ or $\Delta \mathbf{t}$ and $\boldsymbol{v}$ (the groundwater velocity) on the stability of the numerical model.

C. What are the advantages and disadvantages of using numerical methods to solve this problem compared with the method used in Part 2?

Reference:

P. Domenico and F. Schwartz, *Physical and Chemical Hydrogeology*, Wiley, New York, 1990.

Title: Contaminant Transport

Notes for the Instructor

APPLICABLE COURSES. The projects described in this section are currently used at West Point in two courses. The first is a mathematics course that introduces vector calculus and partial differential equations. The second is an environmental engineering course in hydrogeology. Portions of this project can be used in:

1. Multivariable calculus courses that include vector calculus. Evaluation of line and surface integrals is required in the project, as are utilization and interpretation of the theorems of Green, Gauss and Stokes.

2. Differential equations courses that include an introduction to partial differential equations. The method of separation of variables (Fourier's method) is used to obtain an analytical solution of the one-dimensional diffusion equation.

3. Numerical methods courses that include numerical solution of partial differential equations. Explicit finite difference methods are one technique used to solve the diffusion equation numerically.

4. Environmental engineering courses that include hydrogeology as a major component. Computer simulations are used to conduct sensitivity analysis and interpretation of scenarios involving groundwater flow and contaminant transport phenomena.

The engineering course at West Point uses a case study based on a situation at the Rocky Mountain Arsenal and incorporates a particular computer simulation package SUTRA. Copies can be obtained from the U.S. Geological Survey, Open-File Services Section, Western Distributions Branch, Box 25425 Federal Center, Denver, Colorado 80225. The requirements of the engineering design problem are included in the next pages as Parts 4 and 5. They demonstrate the interdisciplinary progression from the mathematics course to the engineering course. Recurring themes in both courses include mathematical modeling, consideration of boundary conditions, the nature of solutions, graphical representation of solutions, and sensitivity of the models to modifications of the parameters.

STUDENT INTERACTION

Mathematics component (Parts 1-3).

Students complete the project in small groups of 2 or 3. The project is given to students at the beginning of the semester with due dates for the various parts annotated in the course syllabus. The timing of the due dates is consistent with course coverage of the required mathematical concepts. For example, Part 1 is due after the coverage of the integral theorems of vector calculus, serving as a review and application of material previously learned. Reference to the project is made by the instructor at relevant times in classroom discussions to motivate discusssions of the topic of the day and to stimulate timely work on the project. Computer algebra systems are used to perform symbolic calculus operations and to plot solutions and results. In Part 3, a spreadsheet program or computer algebra system can be used to obtain the numerical approximations to the solution. The work in the mathematics course is focused on preparing the engineering students to understand the mathematical language and models used in the subsequent engineering courses. Additionally, learning fundamental concepts in the use of numerical methods serves as a foundation to understanding computer simulation of more complex groundwater problems in the engineering course.

Engineering component (Parts 4-5).

These parts of the ILAP need the specific computer code SUTRA. These two parts are included to show the scope of the problem in the engineering course and for use by courses using SUTRA or an equivalent package.

Students should complete the design project in small groups of 2 or 3. A laboratory session is devoted to explain the use of the software package SUTRA and to demonstrate the advantages of computer models over traditional graphical and analytical methods. A specific scenario (Rocky Mountain Arsenal) is used in the course, but can be replaced with any scenario, real or fictitious, which describes the engineering processes and concepts relevant to course learning objectives. Emphasis is placed on design considerations, to include decision-making with consideration of ethical issues involving the potentially conflicting perspectives of the federal government and local communities. Information on the SUTRA package can be obtained from the authors.

Title: Contaminant Transport

Engineering Requirements

Part 4.

Prerequisite Skills

6. Perform sensitivity analysis and interpret scenarios using a groundwater flow simulation program.

Groundwater

At this point, you understand much about the flow of groundwater in the saturated regime and which properties of the porous medium affect this flow. To develop an intuitive sense of what impacts groundwater flow in a macroscopic situation, we are going to use the Rocky Mountain Arsenal as a case study.

You will gain familiarity with computer modeling technology by applying equations for groundwater flow through a porous medium using knowledge of a real set of geologic conditions, specifically those at Rocky Mountain Arsenal, Colorado. After gaining familiarity with the groundwater model and situation, you will perform a sensitivity analysis on major parameters impacting groundwater flow. The project will culminate in an open-ended design for water production for a nearby town.

During laboratory sessions, you will analyze the environmental situation at the Rocky Mountain Arsenal. In those labs you will complete a tutorial on the groundwater modeling code **S**aturated **U**nsaturated **TRA**nsport (**SUTRA**), a hybrid finite element, finite difference simulation program. After completing the discretization of an area of the Arsenal using a finite element grid, you will analyze the problem using assigned parameters. You are to work in groups as assigned by the instructor. To be successful in the lab, you must thoroughly read the assignment for the lesson and these instructions to ensure that you understand the complete procedure.

a. Read the article at enclosure 1 prior to the laboratory period.

b. During laboratory session you will complete a tutorial on the Groundwater Modeling System Trainer (GMST) overlay to the **SUTRA** code using the Rocky Mountain Arsenal case study. We will address questions on the situation, discretization, mathematical models, and the application of the GMST to the situation at Rocky Mountain Arsenal. For the last half of the lab, you will run simulations using your assigned parameters for analysis. The output from the runs will provide the data for the analysis that is to be submitted at the next classroom attendance.

c. **Analysis.** As a part of the design problem submission, you will answer the following questions:

1. What are the initial and boundary conditions used for the model? Ensure that you consider all space dimensions in your explanation.

2. What geologic pattern allows for each boundary condition?

3. What are the magnitudes and directions of groundwater flow under the "default" conditions of the model ? To answer this question you must **PLOT** **contours** of **Steady State Pressures (SSP)** as detailed in the model files. The pressures that "**SSP**" refers to are the fluid potential values within the monitoring piezometers in the flow field. As you know, the direction of groundwater flow will be perpendicular to lines of equal fluid potential (head). Keep in mind that steady state occurs when incremental increases in time do not appreciably change the contour lines of head. On a copy of the contour plot of steady state pressures for your particular set of conditions, draw in arrows generally indicating magnitude and direction of flow. If a homogeneous medium is assumed, the magnitude is controlled by the gradient at each location. To picture this gradient, **PLOT** the **Oblique** of your **SSP** plot to see the three-dimensional depiction of the fluid potential. This will give you a qualitative idea of the gradients at each location. Ensure that you provide these plots as enclosures to your memorandum.

4. Compare the **Contour** and **Oblique** solutions to the initial conditions for the Rocky Mountain Arsenal situation to your hand-drawn Head Lab results (the enclosure given during the Head Lab being the head values at all points in the Rocky Mountain Arsenal

initial situation). (Were you correct? - Does the model provide anything that you did not know? etc.)

5. What assumptions were made in the transition from the real situation to the hydrologic model ?

6. For the default situation, conduct a sensitivity analysis of:

   1) Hydraulic Conductivity, (K)
   2) Porosity (n)
   3) Bed Depth (b)
   4) Time (t)

   What do increases or decreases in these variables do to the magnitude and direction of groundwater flow? Why? Support your conclusions with plots of the potentiometric surfaces (**Contours** and/or **Obliques**).

7. Given the standard input of flow from the contamination site (1 cfs), under what conditions does the potentiometric surface tend to build up ? Why ?

8. Which conditions would be sensitive to establishing a steady state flow field? In other words, if you increase the fluid pressures by well injection, for example, which conditions will take more time to reach equilibrium again? Support your answer with model output.

d. **Design submission contents**. Prepare an "executive summary" briefly describing and summarizing the lab and your results. Attach the following enclosures as a minimum:

1. Answers to questions in analysis
2. Magnitudes and directions of initial conditions by **Contour** and **Oblique** plots.
3. Hand-drawn initial conditions from Head Lab
4. Sensitivity analysis:

   Hydraulic Conductivity, K
   Porosity, n
   Bed Depth, b
   Time, t

References:

1. Konikow, L. and D. Thompson (1984)."Groundwater contamination and acquifer reclamation at the Rocky Mountain Arsenal, Colorado." *Studies in Geophysics: Groundwater Contamination.* National Academy Press, Washington, 93-103.

2. Voss, C. (1984). *SUTRA: A Finite-Element Simulation Model for Saturated-Unsaturated, Fluid-Density-Dependent Groundwater Flow with Energy Transport or Chemically Reactive Single-Species Solute Transport.* US Geological Survey, Water-Resources Investigations Report 84-4369.

Part 5.

Prerequisite Skills:

7. Perform sensitivity analysis and interpret scenarios using a contaminant transport simulation program

Contaminant Transport

At this point, you understand much about the flow of groundwater in the saturated regime and the mass transport of contaminants within this flow regime. To develop an intuitive sense of what impacts contaminant transport in a macroscopic situation, we continue work within the Rocky Mountain Arsenal case study.

Complete the following requirements and answer the following questions for each given scenario:

a. **Sensitivity analysis.** Conduct a sensitivity analysis of the contaminant plume to:

1) Concentration (C, given background C is 10 mg/l)
2) Dispersion coefficient
3) Pumping wells

Specifically, what do increases or decreases in these variables do to the concentration and location of contaminant transport? Why ? Support your conclusions with plots and **Obliques** of concentration and/or potentiometric surfaces. Attach this as enclosure one of the design submission.

b. **Remediation design analysis.** Recent SUPERFUND clean-up accomplishments have improved the quality of the Platte River to the point where it meets current primary and secondary drinking water regulations outlined in the Clean Water Act. As a result, the community identified in Part I decides that the use of the surface water source provided by the Platte River is a more economical option for them than the groundwater source you considered. However, the groundwater contamination at the site you considered in part I can impact the quality of the surface water provided by the Platte River. For the purpose of this study, we are concerned with iron (1000 mg/l of Fe^{2+}) leachate from an abandoned waste disposal site which started operations in 1945 to dispose of empty 50 gallon drums. Records indicate that there are 10,000 kg of iron from these drums in the waste disposal site. In this simulation, the waste disposal site is located at the injection well at node 248, where the flow rate is accepted to be 1.0 ft^3/second from previous water budget studies.

(1) What are the initial conditions used for the model describing contaminant transport ? Remember that you are now dealing with the -CON files, or concentration plumes of the contaminant followed. To answer this correctly, you must include a discussion of what the background concentration is and what the value for it is in this situation.

(2) What happens to the plume when you inject water which has less contaminant than the background concentration in the injection well at node 248? (Provide an oblique plot)

(3) How long does it take for the contaminant to reach the Platte River given your set of conditions due to advection alone? (Use average linear velocity and distance to predict, or let the dispersion be zero.)

(4) How long does it take for the contaminant to reach the Platte River given your set of conditions including dispersive actions? Ensure that you provide plots (enclosures) supporting your answer. Select and justify selection of values for longitudinal and transverse dispersion for input into the model.

(5) If the waste disposal site is estimated to be 50 years old, do we have a problem yet? If not, when will we have a problem?

(6) Given that the loading is **CONTINUOUS**, calculate the mass flux through the aquifer cross section and identify the rate of mass flux into the Platte River (mass/time). Assume that the mass flux has reached the river and is at steady state.

(7) Under the initial conditions of the model, what pumping rate is needed to prevent the contaminant from ever reaching the Platte River in concentrations in excess of the background ?

(8) What is the EPA primary/secondary drinking water standard limit for iron (identify your source)? What pumping rate is needed to reduce iron levels below this standard entering the Platte River indefinitely?

(9) What is the volumetric flowrate entering the Platte River for the cross-sectional area of the steady state contaminant plume in the aquifer?

(10) If the river has a flow rate of 250 cfs of 5 mg/l iron, what is the diluted concentration of the iron at the confluence of the contaminant plume and the river. Assume the contaminant plume to be at steady state.

(11) Is there a problem at the water intake for treatment 1 mile below the plume entry to the river? If so, to what level would you have to reduce the concentrations of iron in the contaminant plume so that the diluted water would meet EPA standards?

(12) If litigation agreements are such that you must treat the contaminant plume to meet the drinking water standard upon its entry into the river, what must the pump(out) rate be for the 3 wells (combined) at the base of the hill?

(13) If it costs two dollars per thousand gallons to treat water which is pumped, what kind of annual budget will you require to execute this for your answers to (11) and (12) above ?

(14) What will you do with the effluent water?

(15) Suggest one methodology for the removal of the iron and describe what will you do with the iron laden material.

(16) How long, and at what cost, must pumping continue if you know from written records, that 10,000 pounds of the contaminant were deposited at the site of contamination? Ensure that you detail your assumptions for the concentrations in this estimation. If the site needs to be remediated within 30 years, can you suggest anything that would reduce the time and cost required to accomplish the pump and treat process?

(17) Are there any engineering applications you recommend to go in concert with this "pump and treat" methodology?

(18) What would contaminant properties such as sorption and/or chemical or biological reactions mean for your predictions of the contaminant plume?

c. **Design submission contents.** Prepare an executive summary describing your design and the results of the sensitivity analysis. Attach all additional calculations and sketches to the summary as enclosures.

Title: Contaminant Transport

SAMPLE SOLUTION

The following is a partial outline of a possible solution to the course project. The software package Mathcad was used to generate this solution, and the Mathcad equations are apparent in the different font. Using a package like Mathcad permits variation of key parameters between groups, permitting a range of possible solutions.

Part I : Applications of vector calculus

Part Ia(1). Stream model

$N := 50 \quad i := 0..N \quad j := 0..N \quad t_i := i \cdot \frac{1}{N}$

$$\vec{r}(t) = -0.7t\,\hat{i} + 5.0t^2\,\hat{j} - 0.245t\,\hat{k}$$

$$X_{i,j} := -0.7 \cdot t_i \quad Y_{i,j} := 5.0 \cdot \left(t_i\right)^2 \quad Z_{i,j} := -0.245 t_i$$

Stream Bed Model

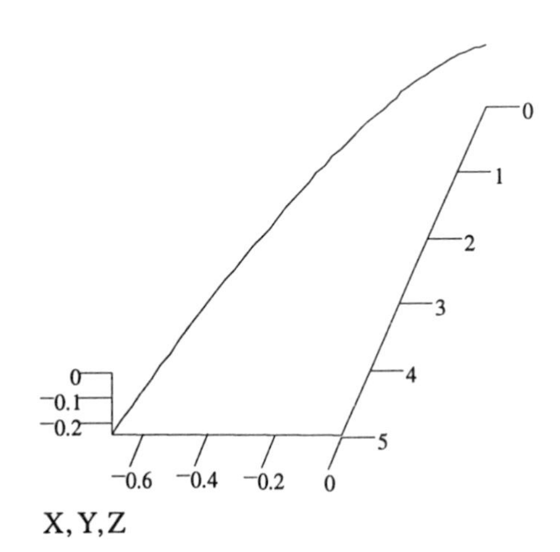

This model is a simplification of the shape of a stream-bed, but has the correct downward path from the landfill to the town.

Part Ia(2). Flow calculations (with units):

$\rho := 1000\frac{kg}{m^3} \quad s := sec$

$x(t) := -700 \cdot t \cdot m \quad y(t) := 5000 t^2 \cdot m \quad z(t) := -245 \cdot t \cdot m \qquad r(t) := \begin{pmatrix} x(t) \\ y(t) \\ z(t) \end{pmatrix}$

$v1(t) := 0.01 \cdot x(t) \cdot \frac{m}{s} \quad v2(t) := -0.005 z(t) \cdot \frac{m}{s} \quad v3(t) := -0.02 \cdot y(t) \cdot \frac{m}{s} \qquad v(t) := \begin{pmatrix} v1(t) \\ v2(t) \\ v3(t) \end{pmatrix}$

$vdotdr(t) := v1(t) \cdot \frac{d}{dt} x(t) + v2(t) \cdot \frac{d}{dt} y(t) + v3(t) \cdot \frac{d}{dt} z(t)$

The flow is the line integral $\int_C F \cdot dr = \int_a^b F(x(t), y(t), z(t)) \cdot \frac{d\, r(t)}{dt}$

$$\rho \cdot \int_0^1 vdotdr(t)\, dt = 1.47 \cdot 10^7 \cdot kg \cdot sec^{-1}$$

This is the mass flow rate of the stream, a positive number consistent with the downhill flow. The magnitude is questionable. The difficulty is due to the scaling of the time parameter, a subject for further discussion.

Part Ib. Flux through the bore holes:

To calculate the flux through the cylindrical surface, Gauss's divergence theorem $\iint_\Sigma \vec{v} \cdot N\, d\sigma = \iiint_V \nabla \cdot \vec{v}\, dV$ can be applied, where S is the cylindrical surface, V is enclosed volume, and the flow field is $\vec{v} = -x\hat{i} - y\hat{j} + 3z\hat{k}$.

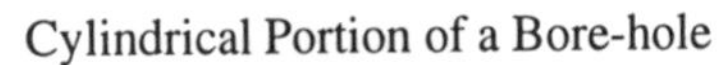
Cylindrical Portion of a Bore-hole

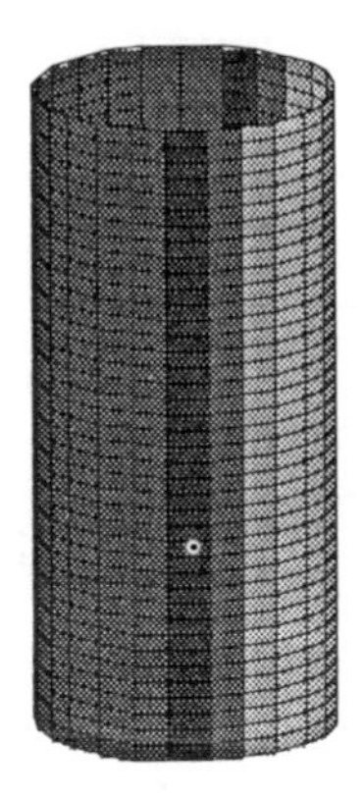

X, Y, Z

The volume integral $\iiint\limits_V \nabla \cdot \vec{v}\, dV$ is computed

$v1(x,y,z) \equiv -x \qquad v2(x,y,z) \equiv -y \qquad v3(x,y,z) \equiv 3 \cdot z$

$$DIV(x,y,z) \equiv \frac{d}{dx} v1(x,y,z) + \frac{d}{dy} v2(x,y,z) + \frac{d}{dz} v3(x,y,z)$$

$DIV(x,y,z) \to 1 \qquad\qquad DIV := 1$

$$\rho \cdot \int_0^{2\cdot\pi} \int_0^{0.1} \int_0^{2} DIV \cdot r \, dz \, dr \, d\theta = 62.832$$

The positive divergence means there is a net fluid flow out of the cylindrical volume. A possible explanation is that water is being pumped out of the borehole.

Note: In this instance, the divergence was a constant, and so the volume of the portion of the well could have been computed using the volume formula for a cylinder. For variable divergence, evaluation of the volume integral will be necessary.

Part Ic. Circulation in the test wells:

The circulation can be determined with the line integral

$$\int_C v \cdot dr = \int_a^b v(x(t), y(t)) \cdot \frac{d\,r(t)}{dt}$$

$x(t) := 0.1 \cdot \cos(t) \quad y(t) := 0.1 \cdot \sin(t)$

$v1(t) := y(t) \qquad v2(t) := -1 \cdot x(t)$

$vdotdr(t) := v1(t) \cdot \frac{d}{dt}x(t) + v2(t) \cdot \frac{d}{dt}y(t)$

$\rho := 1000 \quad \omega := 0.25$

$\rho \cdot \omega \cdot \int_0^{2\cdot\pi} vdotdr(t)\, dt = -15.708$

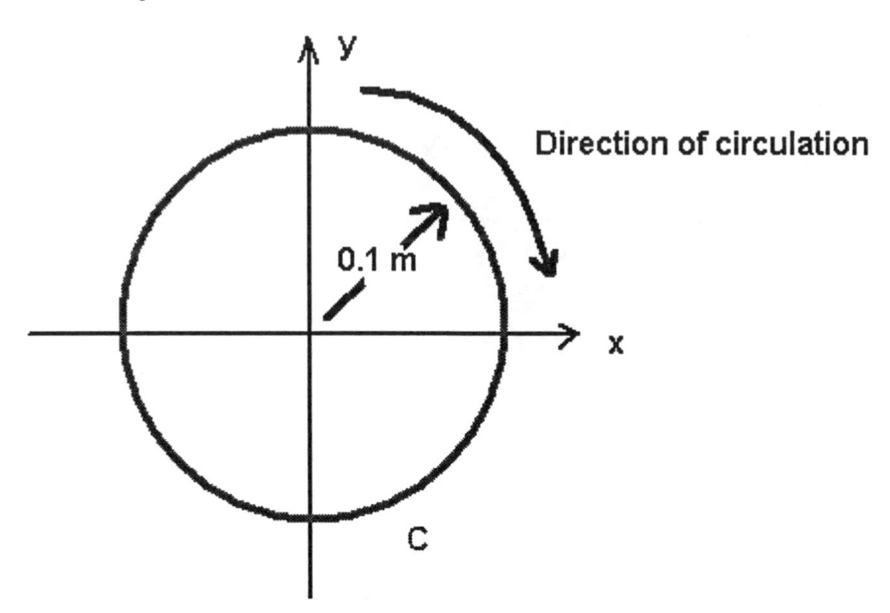

which is above the allowable limits, so the measurements are corrupted by the circulation.

Green's Theorem (or Stokes's Theorem) can also be used to obtain the same result:

$$\oint_C \vec{v} \cdot d\vec{r} = \iint_R \frac{\partial v_2}{\partial x} - \frac{\partial v_1}{\partial y} dA$$

Evaluated symbolically, $\frac{d}{dx}(-x) - \frac{d}{dy}(y)$ is -2, and the area integral over R is

$$\omega \cdot \rho \cdot \int_0^{2\cdot\pi} \int_0^{0.1} -2 \cdot r \, dr \, d\theta = -15.708$$

which agrees with the previous result.

Note: It was not necessary to integrate, as the integrand was a constant and the result can be obtained using the area of the disk, but, for less "nice" problems, evaluating the area integral will be necessary

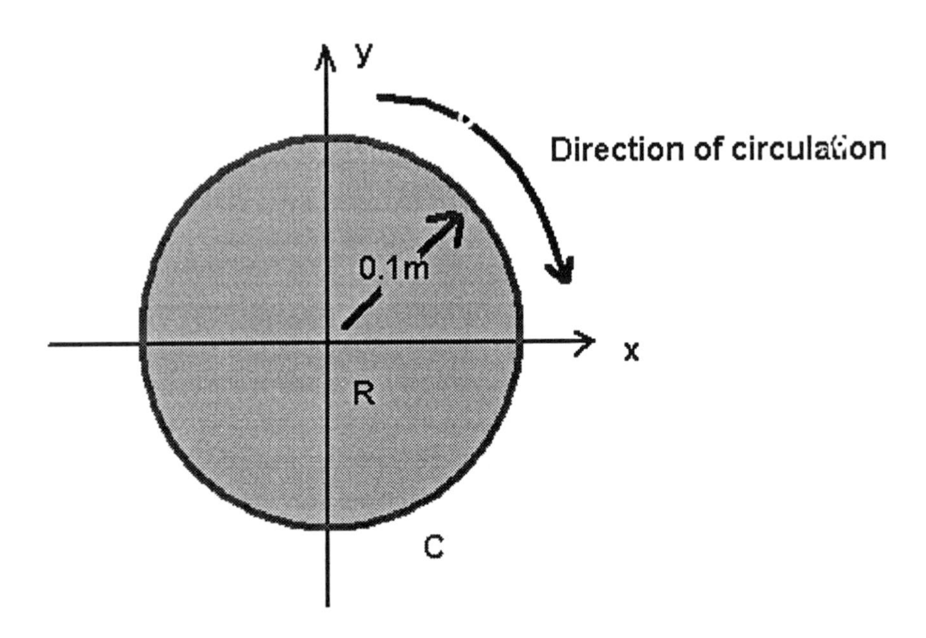

Using Stokes's Theorem, $\oint_C \vec{v} \cdot d\vec{r} = \iint_\Sigma (\nabla \times \vec{v}) \cdot \vec{n} \, d\sigma$ the same integral occurs as the surface **S** is the disk in the xy-plane, with unit normal vector **k**, and curl(**v**) is -2.

Part 2. Analytical solution of the PDE:

First, compute the Fourier coefficients:

$$\frac{1}{5}\cdot\left[\int_0^1 0\cdot\cos\left(\frac{2\cdot n-1}{20}\cdot\pi\cdot x\right)dx+\int_1^4 (x-1)\cdot\cos\left(\frac{2\cdot n-1}{20}\cdot\pi\cdot x\right)dx\right]$$

$$+\frac{1}{5}\cdot\int_4^6 \left(6-\frac{3}{4}\cdot x\right)\cdot\cos\left(\frac{2\cdot n-1}{20}\cdot\pi\cdot x\right)dx+\int_6^{10} \left(\frac{15}{4}-\frac{3}{8}\cdot x\right)\cdot\cos\left(\frac{2\cdot n-1}{20}\cdot\pi\cdot x\right)dx$$

Simplified:

$$10\cdot\frac{\left(14\cdot\sin\left(\frac{2}{5}\cdot\pi\cdot n+\frac{3}{10}\cdot\pi\right)-8\cdot\sin\left(\frac{1}{10}\cdot\pi\cdot n+\frac{9}{20}\cdot\pi\right)-3\cdot\sin\left(\frac{3}{5}\cdot\pi\cdot n+\frac{1}{5}\cdot\pi\right)-3\cdot\sin(\pi\cdot n)\right)}{\left[\pi^2\cdot\left(4\cdot n^2-4\cdot n+1\right)\right]}$$

Since *n* is an integer, the sine terms are zero, leaving

$$10\cdot\frac{\left(14\cdot\sin\left(\frac{2}{5}\cdot\pi\cdot n+\frac{3}{10}\cdot\pi\right)-8\cdot\sin\left(\frac{1}{10}\cdot\pi\cdot n+\frac{9}{20}\cdot\pi\right)-3\cdot\sin\left(\frac{3}{5}\cdot\pi\cdot n+\frac{1}{5}\cdot\pi\right)\right)}{\left[\pi^2\cdot\left(4\cdot n^2-4\cdot n+1\right)\right]}$$

Then the Fourier series approximation is

$x := -10, -9.99 .. 10 \quad m := 15$

$$A(n) := 10\cdot\left[\frac{\left[\left(14\cdot\sin\left(\frac{2}{5}\cdot\pi\cdot n+\frac{3}{10}\cdot\pi\right)-8\cdot\sin\left(\frac{1}{10}\cdot\pi\cdot n+\frac{9}{20}\cdot\pi\right)\right)-3\cdot\sin\left(\frac{3}{5}\cdot\pi\cdot n+\frac{1}{5}\cdot\pi\right)\right]}{\pi^2\cdot\left(4\cdot n^2-4\cdot n+1\right)}\right]$$

$$B(n) := \cos\left(\frac{2\cdot n-1}{20}\cdot\pi\cdot x\right)$$

$$f(x) := \sum_{n=1}^{m} A(n)\cdot B(n)$$

A plot of one period of the (even) Fourier series

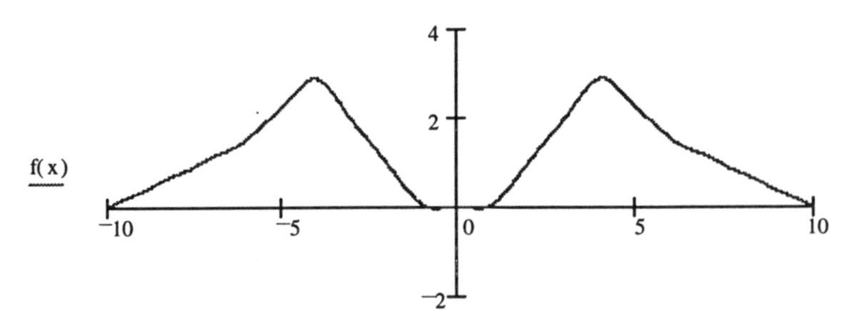

Here's the initial condition in Mathcad terms:

$c1(x) := 0$ $\qquad$ $x1 := 0, 0.1 .. 1$

$c2(x) := x - 1$ $\qquad$ $x2 := 1, 1.1 .. 4$

$c3(x) := 6 - \frac{3}{4} \cdot x$ $\qquad$ $x3 := 4, 4.1 .. 6$

$c4(x) := \frac{15}{4} - \frac{3}{8} \cdot x$ $\qquad$ $x4 := 6, 6.1 .. 10$

Just showing the region from 1 to 10 km, the 15-term Fourier series matches well:

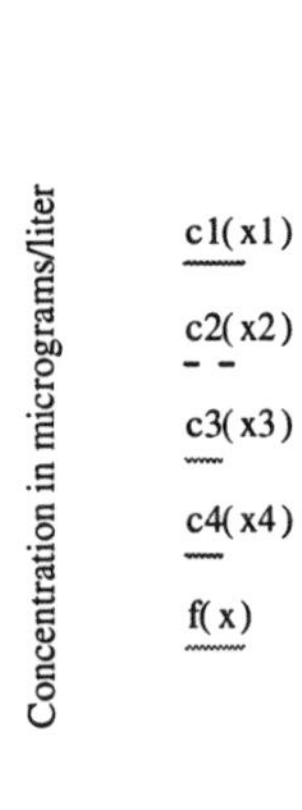

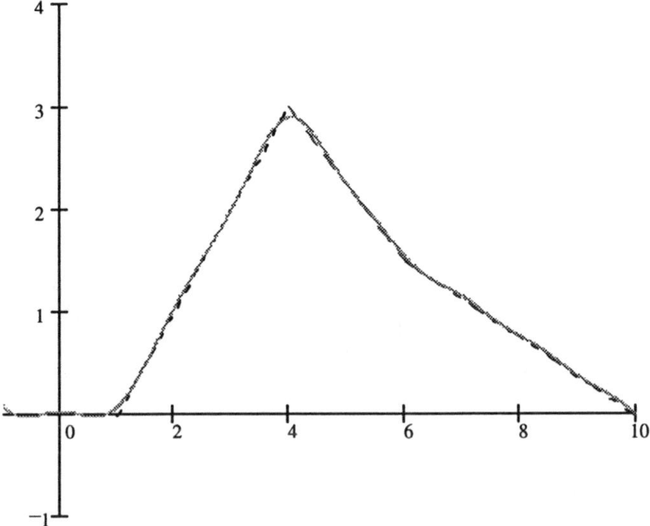

Adding the exponential component:

$$C(x,t) := \sum_{n=1}^{m} 10 \cdot \frac{\left(14\cdot\sin\left(\frac{2}{5}\cdot\pi\cdot n + \frac{3}{10}\cdot\pi\right) - 8\cdot\sin\left(\frac{1}{10}\cdot\pi\cdot n + \frac{9}{20}\cdot\pi\right) - 3\cdot\sin\left(\frac{3}{5}\cdot\pi\cdot n + \frac{1}{5}\cdot\pi\right)\right)}{\left[\pi^2\cdot\left(4\cdot n^2 - 4\cdot n + 1\right)\right]}$$

$$*\cos\left(\frac{2\cdot n - 1}{20}\cdot\pi\cdot x\right)\cdot\exp\left[-0.912\left(\frac{2\cdot n - 1}{20}\cdot\pi\right)^2\cdot t\right]$$

C(x,t)

$i := 0..20$ $\quad j := 0..50$ $\quad x := 0, 0.1..10$ $\quad x0 := 0.0$ $\quad t0 := 0$

$h := 0.5$ $\quad k := 1$ $\quad x_i := x0 + i\cdot h$ $\quad t_j := t0 + j\cdot k$

$M_{i,j} := C\left(x_i, t_j\right)$

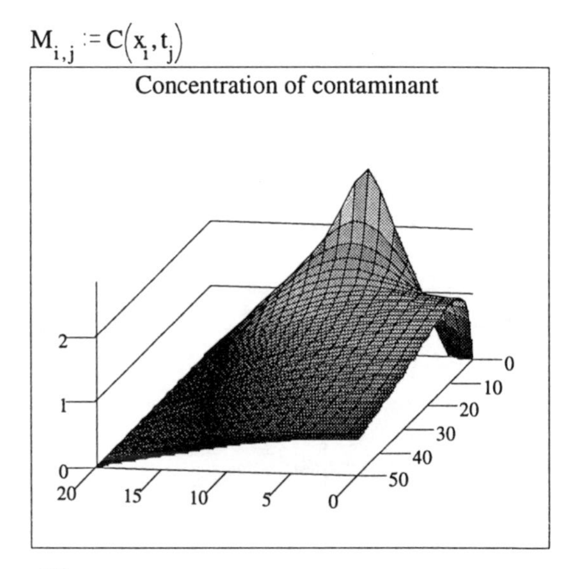

M

The concentration at 5km after 1 year: $C(5,1) = 2.003$ micrograms/liter

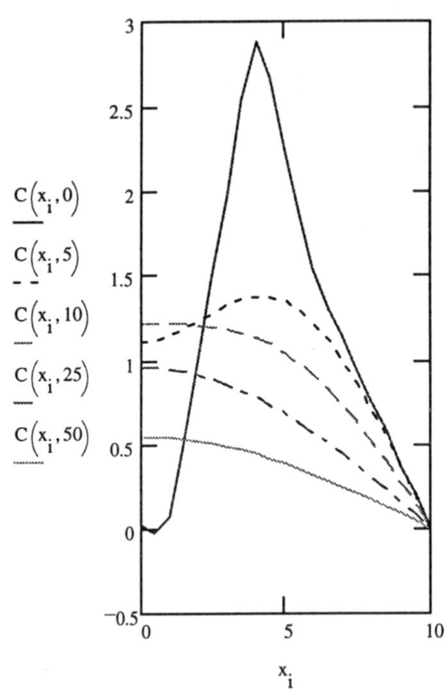

The plots above show the diffusion of the contaminant over time.

Part 3. Numerical solution with and without advection:

Diffusion coefficient: $D \equiv .292$

$$g(x) \equiv \text{if}(x<1, 0, x-1) \quad h(x) \equiv \text{if}\left(x>6, \frac{15}{4} - \frac{3}{8}\cdot x, 6 - \frac{3}{4}\cdot x\right)$$

Concentration at t = 0:

$$f(x) \equiv \text{if}(x \le 4, g(x), h(x))$$

x-step (km): $h \equiv 1$

For stability, k should be less than or equal to: $\frac{h^2}{2\cdot D} = 1.712$

Time step (years): $k \equiv 1$

Number of time steps: $T \equiv 50$ x-steps: $X \equiv 10$ $i \equiv 0..X$ $j \equiv 0..T$

Interior points x_{ii}

$$ii \equiv 1..X-1 \qquad \lambda \equiv \frac{k\cdot D}{h^2} \qquad x_{ii} \equiv ii\cdot h \quad x_i \equiv i\cdot h$$

Initial condition: $u_{0,i} \equiv f(x_i)$

Uses a forward-time, centered-space explicit scheme, no advection.

$$u_{j+1,ii} \equiv u_{j,ii} + \lambda \cdot \left(u_{j,ii-1} - 2 \cdot u_{j,ii} + u_{j,ii+1}\right),$$

End conditions: $u_{j,0} \equiv u_{j,1},\ u_{j,X} \equiv 0$

Plots of the numerical solution, without advection

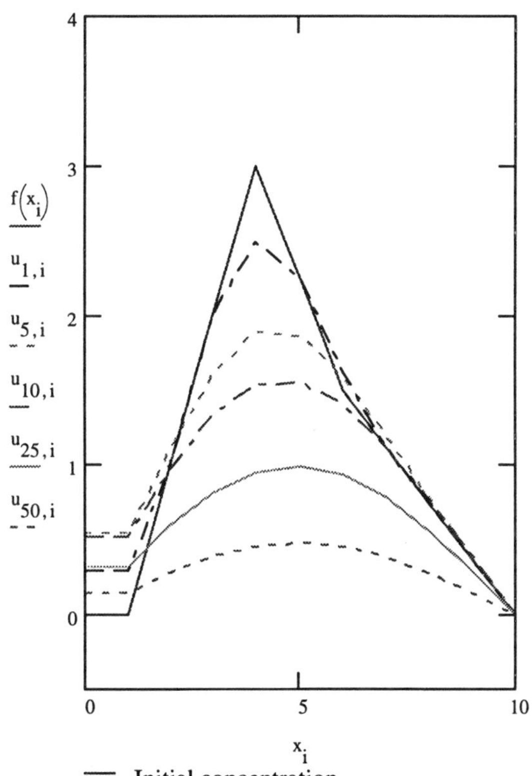

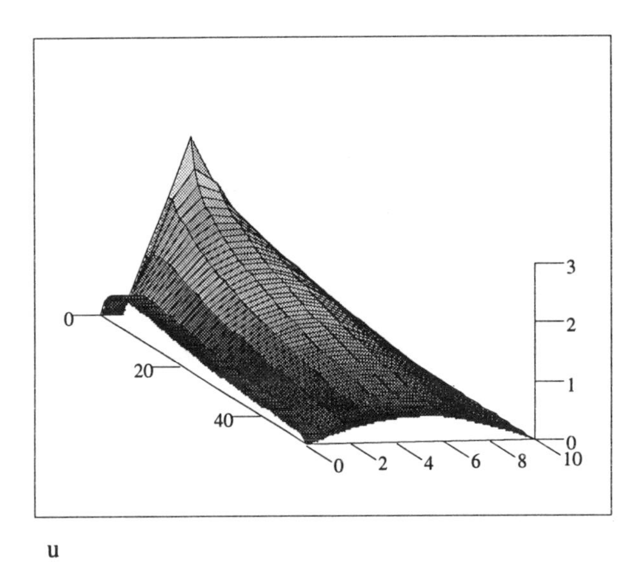

u

Numerical solution with advection:

$$v := 0.1 \quad \mu := v \cdot \frac{k}{2 \cdot h}$$

$$u1_{0,i} := f(x_i)$$

$$u1_{j+1,ii} := \left[u1_{j,ii} + \lambda \cdot \left(u1_{j,ii-1} - 2 \cdot u1_{j,ii} + u1_{j,ii+1} \right) \right] - \mu \cdot \left(u1_{j,ii+1} - u1_{j,ii-1} \right)$$

$$u1_{j,0} := u1_{j,1} \qquad u1_{j,X} := 0$$

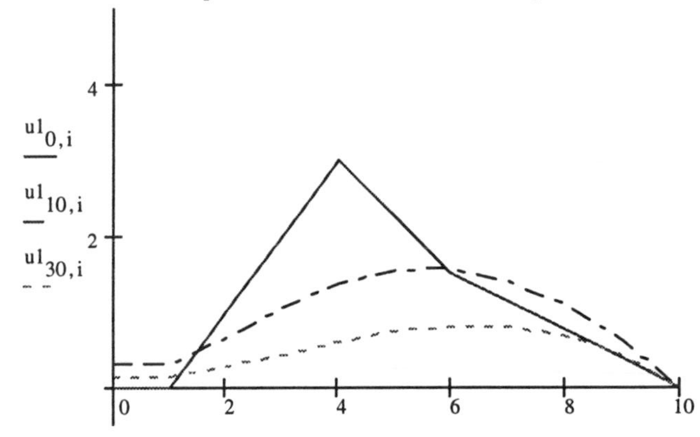

Diffusion of Contaminant

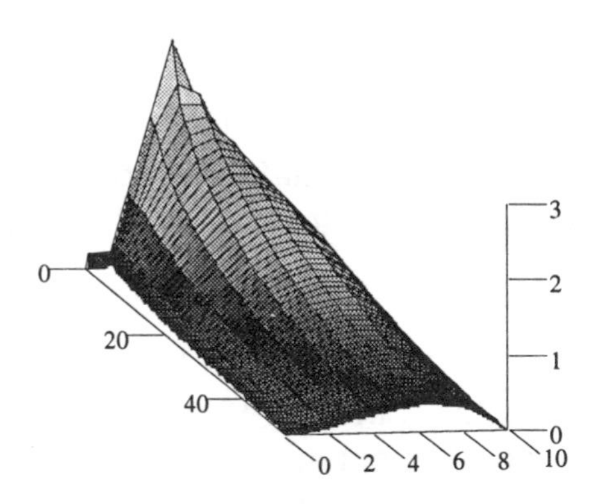

u1

The effect of advection is seen in these contour plots

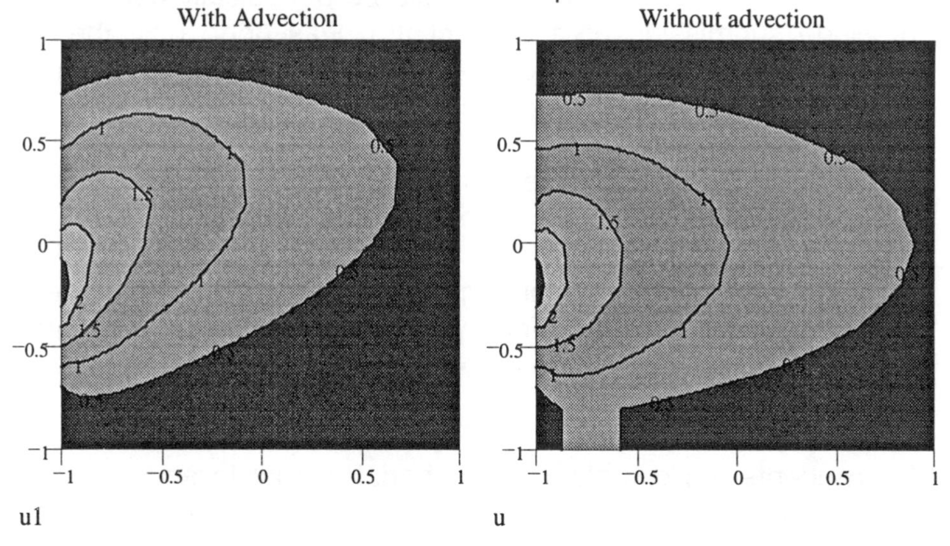

Title: Contaminant Transport

Notes for the Instructor (Student Work on Group Projects)

This project is designed to be completed by students working together in groups. The collaborative process is difficult, but facilitates learning. To increase the effectiveness of the group work, the instructor may implement some of the following activities.

1. Refer frequently to the project in the course of classroom instruction where relevant. For example, in the process of covering applications of line integrals, it is useful to point out to students that the flow integral computation required in Part 1 of the project is now within their grasp.

2. In progress reviews (IPRs) should be periodically conducted to preclude submission of a product that was done only the night before. These can be graded, and usually provide the instructor some insight into the group dynamic.

3. Groups of 4 are too large. Groups of 2 or 3 work best, ensuring more active participation on the part of each student.

4. One way of encouraging group participation is to have the students submit a form signed by all group members indicating the individual contribution to the product. That information can be used to help assign grades.

5. Students will complain that schedule conflicts and geography preclude group work outside of the classroom. There is certainly truth to that; however, the benefits gained in learning from each other is worth the effort for them to work through those difficulties with prior planning.

Technical Report Format and Writing Guide

Kent M. Miller
United States Military Academy
West Point, NY

It is important that mathematicians, engineers, and scientists be able to communicate their technical results effectively in writing. To this end, many colleges and universities incorporate a writing "thread" throughout the curriculum, including science and engineering courses. In mathematics courses, student project reports can form the capstone of the written communication thread. As an integral part of an ILAP project, students can be required to produce a technical report that effectively communicates their analysis and findings.

The purpose of a technical report is to provide the intended audience with a readable, organized presentation of their solution and solution method to a problem. The intent of this requirement for an ILAP is to develop the student's ability to effectively communicate solutions to technical problems. Using a consistent, well organized format can help the development of the student's writing abilities. As a model to follow for ILAP reports, I recommend that the technical report consist of a title page, executive summary, appendices, annexes (as appropriate), and acknowledgments. The outline on the next three pages provides a format guide for the student to use for a technical report. I recommend giving copies of the guide to the students and asking them to follow it when writing their ILAP project technical reports. Use the guide when grading the project to ensure that proper format, content, and writing style are used. For effective feedback, comments to the student should be made using the terminology of the writing guide. Instructors may want to supplement this guide with more details, as needed by their students.

Effective technical writing is not easy, and students need help to develop their skills through practice, experience, and feedback. The use of format guides and classroom discussions can be very helpful in developing student skills and in making the reading of their reports more enjoyable. The following guide is provided for student use:

I. Title Page

II. Executive Summary

The purpose of the executive summary is to briefly describe the problem and to summarize the solution(s) and solution method(s). The intent is for the student to distill the essence of the problem into a document that conveys the information in a clear and concise manner. The summary should be a "stand alone" document, no more than two pages in length. The instructor should evaluate the executive summary in terms of style, substance, organization, and correctness.

Substance

A. Briefly describe the problem and specify the context in which it occurs.

B. Summarize the approach.

C. Summarize the results obtained, both negative and positive as appropriate. Comment on the significance of these results.

D. State briefly whether this problem warrants further examination. If so, recommend the direction future research should take.

Organization

The executive summary should have a logical flow. Start by describing the problem in the introductory paragraph. In the case of a single requirement problem, a brief discussion of your findings is also appropriate in the introduction. In the case of a multiple requirement problem, address the solution to each requirement in separate subparagraphs.

Style

Avoid unnecessary use of the passive voice, slang, technical jargon, and acronyms. The use of simple sentences usually communicates the information best.

Correctness

The document should be free of spelling, grammatical, and punctuation errors.

III. Appendices

Include one appendix for each major requirement or subsection of the problem.

A. Problem Statement

The first paragraph is a concise statement of the problem.

B. Facts Bearing on the Problem

This section contains statements of undeniable facts having influence on the problem or its solution. Exercise care to exclude unnecessary facts that may confuse the issue. Some facts are uncovered during the research and problem-solving stages, while others are inherent in the directive assigning the problem.

C. Assumptions

State any assumptions necessary to solve the problem. Assumptions are used and needed in the absence of factual data. The assumption, while not a fact, must have a basis in fact. Justify your assumptions.

D. Analysis

This section presents a detailed solution to the problem. Don't just write line after line of mathematical equations. Embed the mathematics and symbols in coherent sentences. Omit any of the following sections that are not applicable.

1. Definition of Variables and Symbols

2. Methodology

3. Formulas and Expressions Used and Manipulated

4. Calculations

5. Diagrams (Essential Plots and Graphs)

6. Discussion of Results

7. References to Supported Annexes

E. Conclusions

Conclusions must follow logically from the analysis. Do not introduce new material in this section. Answer the problem statement directly.

IV. Annexes

A. Other Supporting Plots and Graphs

B. Computer Output

C. Data Lists

V. Acknowledgments

Project INTERMATH: An Interdisciplinary Approach to Cultural Change

Richard D. West
United States Military Academy
West Point, NY

Introduction

In 1993 the National Science Foundation called for a program of interdisciplinary activities to effect cultural change in undergraduate mathematics-based disciplines across the nation. Project INTERMATH is one of seven major initiatives funded to date by the NSF under this program. Its purpose is to inspire mathematics and the disciplines it supports to continue in the spirit of the Calculus Reform movement in effecting cultural and curricular change.

Project INTERMATH is a consortium of 12 schools led by the United States Military Academy at West Point. Interdisciplinary activities included in this initiative are centered around the process and use of Interdisciplinary Lively Application Projects (ILAPs), small-group projects developed by faculty and experts from more than one discipline. Our plan is to promote reform through ILAP production, curriculum design, and conferences and workshops. We believe that the process of developing and the use of these ILAPs generate the communication, involvement, and connections needed to effect educational change. In this paper, I investigate the core characteristics of the national reform movement, discuss what it means to change culture, and make a case for the use of interdisciplinary projects (ILAPs) as the impetus for change.

Reform

In response to a national call for reform in mathematics that began back in the late 1970s, the Committee on the Undergraduate Program in Mathematics (CUPM) issued its *Recommendations for a General Mathematical Sciences Program* (Tucker, A., 1981). The writers of this document called for a move away from the traditional mathematics sequence of courses to an integrated, more applied mathematical science curriculum (Tucker, A., 1981). Although educators appeared ready to solve many of the issues of the times, there was no consensus in the mathematics community about how to proceed, and the 1981 CUPM report generated little change. In subsequent reports (CBMS, 1984; NRC, 1984; National Science Board Commission on Precollege Education in Mathematics, Science and Technology, 1983; USDE, 1983), educators further documented deficiencies in the mathematics curriculum at all levels of schooling and called for similar changes to be made at the precollege level as well.

Not until the publication of additional, more urgent cries of crisis (Douglas, 1986; McKnight et al., 1987; National Council of Teachers of Mathematics, 1989; NRC, 1989; Steen, 1989) did many of the earlier recommendations get acted upon. The curriculum goals addressed in both of these documents include content goals of integrating the topics of discrete mathematics and probability and statistics throughout the curriculum, and instructional goals of integrating the use of computer and calculator technologies as tools for instruction and learning. They also encourage the use of writing and mathematical modeling throughout the curriculum. In addition, college-level mathematical sciences programs require a smoother transition from secondary to collegiate mathematics and a less special case and topic-ridden and more technology and applications-oriented calculus. A major objective of the reform effort is to prepare students to apply mathematics to problem-solving situations, to motivate them to seek further learning, and to enable students to handle the technical and complex problems of the applied sciences, engineering, and social sciences.

Over the past few years, the focus of reform at the undergraduate level has been on changing calculus. As a result of a number of initiatives, many reform calculus texts are now on the market, and the use of graphing calculators and computer algebra systems is being integrated into many calculus classrooms. In general, the calculus reform and the national reform movement have shifted the emphasis in the classroom from what we teach to how we teach. This shift could be further characterized as a move from teacher-centered to student-centered classrooms. The focus is on students' learning and development.

Reform has taken diverse forms at different institutions. For some, changing the sequence of courses is called reform. For others, changing to a newer text may be called reform. Further, technology has added lab components at many institutions. And others have integrated math modeling and applications throughout their curriculum. The measures of success of these implementations would be: how effective was the reform in improving student learning?

Cultural Change

If we are to respond to the call for change from the national mathematics reform movement, we need to change the way we teach and communicate. Furthermore, to accomplish cultural change, the goals are best developed at and for the institution. These goals should be focused on "better" (more efficient and longer lasting) student learning and should be generated by the faculty who teach these students.

Faculty members who are involved in this kind of change have the sense that the change occurred because it was needed and the reform helps the students learn better. Further, students and faculty benefit: in the enhancement of the relevance of mathematics in a better connected and coordinated curriculum; in opportunities for extensions beyond the current content; and from a comprehensive educational experience.

ILAPs

Interdisciplinary Lively Application Projects are projects developed with partner disciplines. Students benefit from small-group projects, because the students work together to formulate and solve problems rather than being compared individually against a standard. Communication is obviously enhanced through group work. Students feel more open to trying ideas. Mathematical modeling and scientific problem solving are two of the hardest tasks for students to learn, and the group gives a good forum for evaluating and improving on informed conjectures. Further, ILAPs often encourage the use of computer or calculator technology in the modeling, solution, and presentation processes. Finally and most important, these projects are opportunities for students to construct their own mathematics through discovery learning.

The development of ILAPs is an enabling process for the faculty involved in the design and implementation of these projects. Faculty communicate with members of other disciplines in the design process. They generate pedagogical ideas together in the implementation planning process. This enthusiasm for the projects through application and ownership is carried into the classroom where students can benefit. It is our experience that, once faculty members have participated in the process of developing an ILAP and used it in the classroom, they are more willing to use projects produced by others. Further, once instructors see the benefits to student learning from these interdisciplinary projects, they want to use them more, and there is a shift of emphasis in the classroom from content coverage to student-centered learning. In short, faculty members change the way they teach. The resultant open communication across disciplines, the desire to connect mathematics to other disciplines, and the changes in the way we teach accomplish over time our desired cultural change.

Project INTERMATH

At West Point, we have seen ILAPs bringing about cultural change and accomplishing our goals. Our students' communication, reasoning, and modeling skills have been enhanced. The students' use of technology has enabled the tackling of difficult problems. Students learn to value mathematics as a human endeavor used for solving real world problems. Further, through the projects we can connect to mathematics learned in the past, extend to mathematics understood more completely in the future, and connect to other disciplines. The cultural change at West Point has evolved to where we are now ready to continue our process of change and disseminate our findings to others. The scope and history of West Point's activity with projects and ILAPs are given in this volume in another article. We envision cultural change taking place at all of our 12 consortium schools.

Summary

In conclusion, we believe cultural change can be accomplished through the development of interdisciplinary small-group projects (ILAPs) that change the way we teach and eventually change what we teach in the process.

References

Conference Board of the Mathematical Sciences (CBMS). (1984). New goals for mathematical sciences education. Washington, DC: Author.

Douglas, R. (Ed.). (1986). Toward a lean and lively calculus. MAA Notes Number 6: Mathematical Association of America (MAA).

McKnight, C., Crosswhite, F., Dossey, J., Kifer, E., Swafford, J., Travers, K., & Cooney, T. (1987). The underachieving curriculum: Assessing U.S. school mathematics from an international perspective. Champaign, IL: Stipes Publishing.

National Council of Teachers of Mathematics (NCTM). (1989). Curriculum and evaluation standards for school mathematics. Reston, VA: National Council of Teachers of Mathematics.

National Research Council (NRC). (1984). Renewing U.S. mathematics: Critical resource for the future. Washington, DC: National Academy Press.

National Research Council (NRC). (1989). Everybody counts: A report to the nation on the future of mathematics education. Washington, DC: National Academy Press.

National Science Board Commission on Precollege Education in Mathematics, Science and Technology. (1983). Educating Americans for the 21st Century. Washington, DC: National Science Foundation.

Steen, L. A. (Ed.). (1989). Reshaping college mathematics. MAA Notes Number 13: Mathematical Association of America (MAA).

Tucker, A. (Ed.). (1981). Recommendations for a general mathematical sciences program: A report of the Committee on the Undergraduate Program in Mathematics. Washington, DC: Mathematical Association of America (MAA).

U. S. Department of Education (USDE), National Commission on Excellence in Education. (1983). A nation at risk: The imperative for educational reform. Washington, DC: US Government Printing Press.

ILAP Products: Authoring, Testing, and Editing

David C. Arney
United States Military Academy
West Point, NY

Introduction

The authoring, editing, production, and dissemination processes for additional Project INTERMATH's ILAP products will be accomplished in a setting of interdisciplinary cooperation and partnership. COMAP will be the publisher for these additional products. Both the processes and structure require individuals from the partner disciplines to work together.

ILAPs can be viewed from several perspectives. From a curricular and pedagogical point of view, ILAPs are tools used to develop quantitative problem solving skills; to motivate and give relevance to student learning; to instill and reinforce the principles of mathematical modeling, scientific method, engineering thought process, and quantitative reasoning; to provide a framework for student growth and curricular improvement; and to link faculty and students from different disciplines. From a product point of view, ILAPs are efficient and effective course material. They begin with background material and a problem statement and are supplemented as needed with an instructor guide, sample solutions, and motivational video or other media presentations. ILAPs are the means by which faculty from different disciplines come together to discuss, design, and implement the curriculum and pedagogy and by which students from different disciplines learn to appreciate one another in a team setting. The authoring, editing, production, and dissemination processes are designed to support effectively all these perspectives of the ILAP products.

We expect to have many faculty and students from many disciplines involved as ILAP users. We also expect to have many faculty from many disciplines involved in the authoring, editing, and testing processes. Some faculty will initiate a product by becoming ILAP authors. Others will be involved as testers, reviewers, refiners, and users. There are several editors representing different disciplines and a Managing Editor to coordinate the editing and testing.

Products:

The following list describes the ILAP products that are being managed and produced. The ILAP projects contained in this volume are "working" projects, ILAP Projects, or ILAP Carry-Through Projects.

"Working" project -- A problem statement for an interdisciplinary project that has been discussed by two individuals from different disciplines. "Working" projects are cataloged and available from the electronic data base. A "working" project may be further developed into other products.

ILAP Project -- A project that has been refined, tested, edited, cataloged, and is ready for dissemination. An ILAP Project usually contains a problem statement, sample solution, and instructor guide. ILAP projects are available in hard copy as well as electronically.

ILAP Super Project -- A large-scale interdisciplinary project that involves several disciplines and a group effort to understand and solve. Super Projects are often very broad and usually require many hours of effort to solve. These projects can be used as major requirements for a course and are probably best used for experienced problem-solvers in a capstone course.

ILAP Carry-Through Project -- An ILAP Project with several stages that build on the same scenario. The later stages usually build from models developed in previous stages. This makes it possible for one scenario to be carried through an entire course or set of courses, enabling students to experience solving problems by model refinement or by use of different perspectives and assumptions.

Production and commercial sale of additional hard-copy ILAP Projects, other than those given in this volume, will be performed by COMAP. COMAP has the publishing experience and new technology to perform interactive publishing that allows individual instructors to customize the project hand-outs for their course. COMAP can be contacted at Suite 210, 57 Bedford Street, Lexington, MA 02173.

Process:

There is no one fixed process to develop ILAP products. However, there are some guidelines and a general flow through stages of development to ensure that the goals of the project are accomplished. The flowchart given in Figure 1 outlines the general process used for the development of an ILAP classroom product. In order to encourage interdisciplinary cooperation, all ILAPs will have joint authorship and the authors must represent at least two different disciplines. There are two editors from different disciplines assigned to each project. The editors will classify the project by subject in mathematics, disciplines of the application, and the requisite mathematical and disciplinary skills.

The classification system would include, but not be limited, by the following lists:

Mathematical subjects: Algebra, Trigonometry, Geometry, Precalculus, Differential Calculus, Integral Calculus, Vector Calculus, Differential Equations, Discrete Mathematics, Probability, Statistics, Linear Algebra, Abstract Algebra, Operations Research, General Mathematics.

Disciplinary subjects: Chemistry, Physics, Mechanics, Biology, Economics, Engineering (including its various subfields), Geography, Topography, Oceanography, Technology, Computer Science, Operations Research, Life Science, Exercise Physiology, Social Science, Medicine.

Figure 1. Flow for the development of an ILAP product.

Conclusion:

The authoring and editorial organization and processes are designed to take place in a setting of interdisciplinary cooperation and partnership. Both the processes and structure require individuals from the partner disciplines to work together to accomplish the authoring and editorial tasks. ILAP projects are much more than just products or tools; they are the means by which faculty and students from different disciplines become educational and problem-solving partners. It is envisioned that many ILAP products will be used in several subjects and levels of undergraduate education.

ILAPs: A Vehicle for Curriculum Reform

Frank R. Giordano
COMAP
Lexington, MA

Introduction

We illustrate how ILAPs connect math courses together to form a more integrated mathematics curriculum. We show how ILAPs ignite a process in which other disciplines become partners in designing the undergraduate experience in math, science, and engineering.

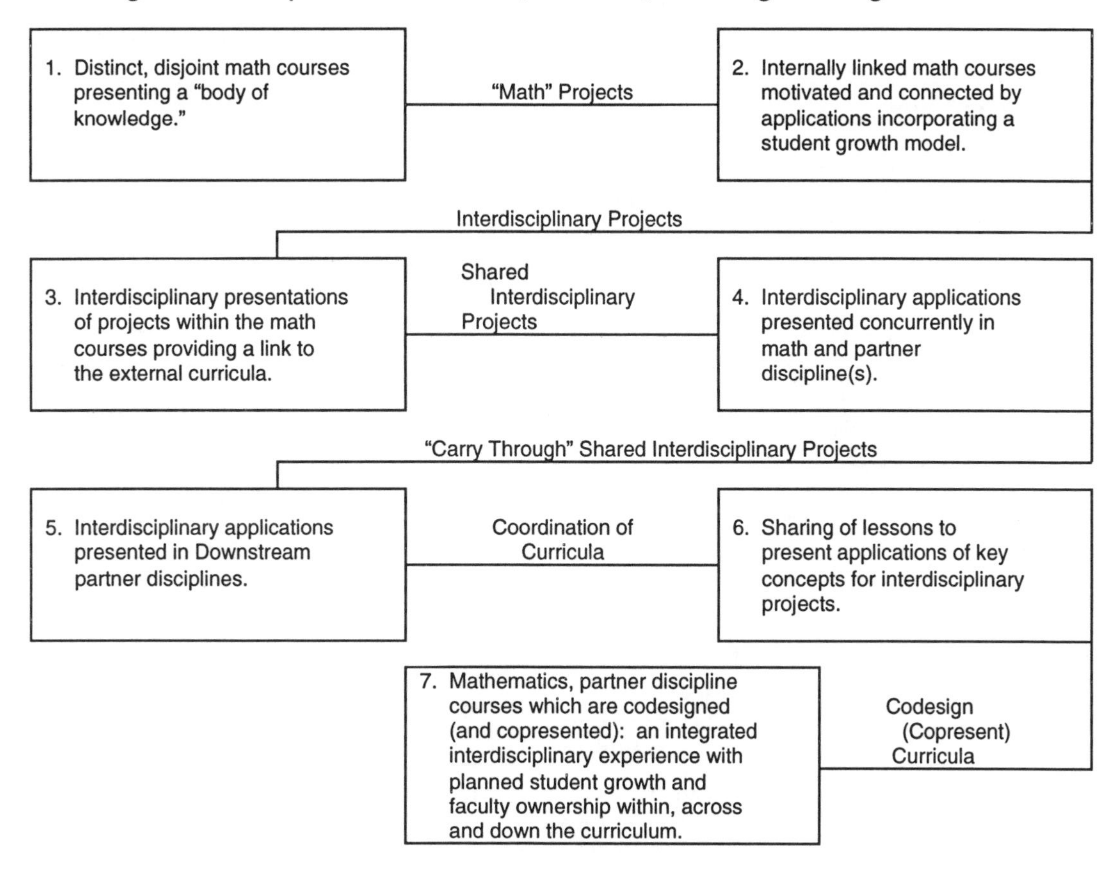

Figure 1: End states (rest points) and processes for achieving integrated interdisciplinary curriculum.

Moving from Disjoint Math Courses to Integrated Math Curricula

In Figure 1 we depict 7 possible end states that describe the state of various mathematics courses and their relationship with the courses of the partner disciplines. Except for the first end state, each end state might be appropriate for the conditions at a particular institution. For example, some institutions may not have a "core" mathematics program that can be "linked" to the partner disciplines. However, for many institutions, especially ones with a "mathematics core", Figure 1 may also be viewed as a model for the evolution of an integrated, interdisciplinary curriculum. In the model a "process" is identified to effect the transition from one end state to the next if evolution is desired. These "processes" are key to Project INTERMATH and provide the focal point for the project. The effect of the processes on student growth, faculty development, and the resulting culture is summarized in Figure 2.

End State	Transition Process	Opportunities for Student Growth	Requirements for Faculty Development	Cultural Change
From: 1 To: 2	Math Projects	Planned Growth in critical development areas (threads)	Knowledge of Applications	Pedagogy; Ownership within Mathematics
From: 2 To: 3	Designing Interdisciplinary Projects	Relevance of Mathematics	Work in Teams	Empowerment
From: 3 To: 4	Sharing Projects	Coordinated Student Growth Threads	Coordinate Curricula	Opens Interdisciplinary Communications
From: 4 To: 5	Extending Projects	Extended Growth	Integrating a Program of Study	Bridge for Connected Curricula
From: 5 To: 6	Coordinating Curricula	More Comprehensive Growth Model	Design Curricula	Ownership across Curriculum
From: 6 To: 7	Codesigning Curricula	Smooth Comprehensive Growth Model	Design a Comprehensive Educational Experience	Integrated Interdisciplinary Curricula

Figure 2: Curriculum reform--the process for cultural change

Processes

Projects offer several opportunities for enhancing both student and faculty growth as well as for affecting curriculums. For example, projects can be designed to provide connections between mathematics courses as well as to motivate and to summarize key concepts. Also projects designed to help students take responsibility for their own learning are powerful instruments for developing student-centered curricula. Effectively incorporating projects into a curriculum often requires a change in pedagogy. In particular, an instructor's role changes from one of presenter to one of facilitator. **(From State 1 to 2)**

Projects whose design and presentation involve faculty from partner disciplines help students realize and appreciate the relevance of mathematics. Participating in such a design team provides a sense of empowerment to individual faculty members as they become involved in shaping the mathematics curriculum. **(From State 2 to 3)**

A next step of sharing projects between concurrent mathematics and partner courses requires a high level of collaboration and coordination. This collaboration between departments, particularly with respect to planning student growth, is often difficult or nonexistent in many campus cultures. **(From State 3 to 4)**

A Process for Moving to Interdisciplinary Experiences

Extending projects to downstream courses which require more sophisticated analysis provides coordination of the curriculum as well as coherence to extended student growth models. **(From State 4 to 5)**

At this point communication and cooperation among partner disciplines have advanced to a state permitting the coordination of curricula across departments resulting in a more comprehensive growth model than a single department could design. Faculty ownership through participation is extended across the curriculum. **(From State 5 to 6)**

Finally, co-designing curricula results in a smooth comprehensive growth model, an important part of an integrated interdisciplinary curriculum. **(From State 6 to 7)**

Summary

ILAPs can be a useful vehicle for starting a process that can lead to integrated and interdisciplinary curricular experiences. We hope that the continued progress of Project INTERMATH has a positive impact on the academic culture of teaching and learning mathematics and on all of its partner disciplines.

Interdisciplinary Communication and Understanding

John H. Grubbs
United States Military Academy
West Point, NY

and

James H. Stith
The Ohio State University
Columbus, OH

"...and they all lived happily ever after."

-Anonymous

Introduction

It would be nice indeed if all the stated goals of Project INTERMATH would just magically happen. They won't. Given the difficult nature of bringing about change, we will present our thoughts about interdisciplinary communication (sharing information) and understanding (appreciating how partner disciplines contribute to one's primary discipline). While we view mathematics as the cornerstone for our efforts, success in Project INTERMATH certainly goes beyond a simple appreciation of the role mathematics plays in education. As a component of the MAA Classroom Resource Volume, this article presents a case study in which communication between the Department of Mathematical Sciences at the United States Military Academy and "client" science and engineering departments has improved. As a matter of fact, the word "client" can now be replaced with "partner". Some interesting data from the United States Military Academy will be presented. The two authors of this article hope to continue our involvement with Project INTERMATH by communicating and disseminating Project INTERMATH ideas on a national scale and using ILAPs in our engineering and science courses.

A History: Genesis of Project INTERMATH

In the late 1980s the leadership of the Department of Mathematical Sciences (please note that neither of the authors is a mathematician) at the United States Military Academy sensed that, while interaction with other departments was very cordial, there was a disconnect in the thread of mathematics as cadets moved from taking core mathematics courses to using mathematics in their upper division disciplinary programs. To many, mathematics was something to be endured. To improve upon this unacceptable situation, the department instituted a series of initiatives that brought about a change in the manner in which mathematics was viewed by others and used by the cadets. Associated goals of this effort include:

- To improve student understanding of the relevance of mathematics to other disciplines.

- To develop a linkage between mathematics and other disciplines -- particularly with science and engineering.

- To improve direct communications with other departments, making them both customers and partners.

- And, through the use of the ILAPs, to change mathematics from being an educational filter to an educational pump.

Without question, the fundamental engine for Project INTERMATH is the ILAP.

Results in Understanding and Communication: Can They be Measured?

In both qualitative and quantitative terms, Project INTERMATH can be categorized as an incredible success story. While many factors combine to produce a first-rate program in any discipline, the positive outcomes of Project INTERMATH at West Point leave no doubt as to its worth to our mathematics-science-engineering curriculum.

Qualitative:

Beyond feeling "warm and fuzzy" about Project INTERMATH, specific qualitative improvements have evolved since 1990. Among them are:

- An increased degree of communication between faculty of the Department of Mathematical Sciences and each of the engineering departments. Similar communication exists with other departments as well. An example of this increased communication has been seen in the designation of a mathematics professor to work with each engineering and science department at the Academy.

- Implementing a briefing program in which the Department Head of the Department of Mathematical Sciences provides an overview of the mathematics program to the engineering and science faculty at the beginning of each academic year.

- A strong sense of collegiality among the Heads of all Mathematics-Science-Engineering (MSE) departments. Although the MSE Committee was in existence prior to the initiation of Project ILAP, an increased sense of "working together" pervades the MSE Committee, which consists of the chairs of the departments.

- An increase in activities centered about developing interdisciplinary programs and ILAPs involving all MSE disciplines.

Quantitative:

Data that can be measured numerically can also be linked to Project INTERMATH. Specific indicators include:

- The percentage of cadets who choose a major or field of study (fourth semester) in mathematics, basic science, or engineering (MSE) has risen since 1990. This trend has continued through the class of 1996 as seen below:

	1990	1996
MSE	49%	58%
Humanities	51%	42%

- In each of the four years following 1990, the Quality Point Average (QPA) for cadets enrolled in the core mathematics curriculum has improved.

- Responses of the class of 1996 to questions pertaining to their experiences at West Point yielded uncommonly positive results. In reporting upon their degree of satisfaction with the Academic Goal, **"To establish a sound foundation in mathematics and the physical sciences,"** 91% stated that they were either "Fully Satisfied" (41%) or "Satisfied" (50%). For the Academic Goal, **"To learn to use the engineering process by which mathematical and scientific facts and principles can serve human purposes,"** 92% stated that they were either "Fully Satisfied" (41%) or "Satisfied" (51%).

Full Interdisciplinary Potential

Whereas the emphasis of this discussion has focused on the MSE disciplines, the success in Project INTERMATH applies to other disciplines as well. The data presented above is not intended to make a case for diverting students away from the humanities. It is, however, very significant in that West Point tends to attract young men and women who excelled in mathematics and the sciences in high school. The fact that many cadets had "lost" interest in majoring in the MSE disciplines was seen as a significant problem related to their experience in the mathematics and science classrooms at the Military Academy. The increased percentage of cadets majoring in MSE disciplines is seen as (1) appropriate, (2) healthy for the Academy, and (3) a consequence of ILAPs and the other aspects of Project INTERMATH.

Full health, however, will become reality only when all potential partner academic departments become involved with using mathematics to enrich their own academic programs.

Summary:

Those of us involved in Project INTERMATH are excited about the future. We would be glad to discuss -- and to receive new ideas concerning -- our goal of changing the culture of the relationship between mathematics and all other disciplines. The vision of Project INTERMATH is one of departments working together to develop interdisciplinary applications demonstrating the utility of mathematics in formulating, investigating, and solving real problems in economics, engineering, and the physical and social sciences. That vision will be fulfilled when we change mathematics from being seen as the proverbial "filter" to being a "pump" for student success. We invite your participation.

References:

1. Grubbs, John H., "Using Environmental Pollution Problems to Improve Mathematics Education - What's an 'ILAP'?", Proceedings of the 1995 Annual Conference, American Society for Engineering Education, Anaheim, California, June 1995.

2. Le Masurier, David. "Pollution of the Great Lakes," Mathematical Modeling - A Source Book of Case Studies, Oxford University Press (1990), pp. 181-193.

3. An Introduction to the Interdisciplinary Lively Applications Project, The Consortium for Mathematics and Its Applications (COMAP), Lexington, Massachusetts, 1995.

Interdisciplinary Projects at West Point

Kelley B. Mohrmann and Richard D. West
United States Military Academy
West Point, NY

Introduction

Since the inception of the United States Military Academy at West Point in 1802, the faculty of its Department of Mathematical Sciences has been interested in applied problems. In the early years of the 19th century, cadets faced the engineering problems associated with building the nation and learned the relevance of mathematics from that experience. As the 21st century approaches, cadets and faculty at West Point still look to explore and explain more complex and evolving challenges.

Faculty at West Point continually cooperate in order to enable cadets to develop confidence and competence in using mathematics as well as an understanding of its importance and utility by investigating problems in various applied settings. The increasing impact of technology on teaching and learning has resulted in even more interdepartmental cooperative efforts. Over time these have become a standard method of developing curricular innovations and have produced many multidisciplinary projects and activities. What follows is a list of some of the more recent examples of such projects and activities. Since many of the projects in this list are military-related and written specifically for cadets, their general applicability may be somewhat limited. Copies of many of the projects listed, as government-produced public domain information, are generally available by writing the authors of this article at the Department of Mathematical Sciences, West Point.

The problems listed as Projects (not ILAPs) are applied problems generated by faculty in the mathematics department only. At the same time, many of these application projects have the potential of becoming ILAPs. Generally, the problem merely needs to be expanded with faculty from a partner discipline. When no affiliation is given for an author, he or she is a member of the Department of Mathematical Sciences, West Point.

ILAP and Project Catalog at West Point

MA100 - PreCalculus

Air Traffic Controller Problem -- Project

Developed by Dave Nadeau in 1993. Later modified by Joe Schulz as a project in Calculus I for the Fall 1993 and later used by Barbra Melendez as a project in Calculus II.

Profit Management of a Pig Farm -- Project

Developed and used as a project in Fall 1995. Developed by Ken Koebberling.

MA103 - Discrete Dynamical Systems and Introduction to Calculus

1D Heat Transfer -- ILAP

Developed and used with Department of Civil and Mechanical Engineering (CME) in Fall 1992. Bob Potter (CME) delivered lecture.

Pollution along a River (the PUKE problem) -- ILAP

Developed and used with Department of Geography and Environmental Engineering (GEnE) in Fall 1992. Chris King (GEnE) delivered lecture. Quattro Pro slide show developed by Craig Russell for Parents' Weekend Open House. Later developed as an in-class activity by Jeff Strickland as a differential equation model in Calculus I in Spring 1992 and Spring 1994.

Tank Battle (direct-fire simulation) -- ILAP

Developed and used with Department of Systems Engineering (SE) in Fall 1992. Mark Tillman (SE) delivered lecture/video. Quattro Pro slide show developed by Craig Russell for Parents' Weekend Open House.

Chemical Chain Reaction -- ILAP

Developed and used with Department of Chemistry in Spring 1993. Stan March (Chem) delivered lecture in the classroom.

MA103 - Discrete Dynamical Systems and Intro to Calc (continued)

Great Lakes Pollution (3 lakes as a single equation) -- ILAP
Developed and used with GEnE in 1993. Chris King (GEnE) delivered lecture. Video done by GEnE. Quattro Pro slide show developed for Parents' Weekend Open House by Alex Heidenberg. COMAP published this ILAP in NSF Sample Book in 1995 as a carry-through problem involving Discrete Dynamical Systems, ODEs, & Probability & Statistics.

SMOG in LA Basin -- ILAP
Developed and used with Department of Chemistry in Fall 1993 & Spring 1994. Chuck Bass (Chem) developed video. Quattro Pro slide show developed for Parents' Weekend Open House by Alex Heidenberg. This ILAP appears in this MAA volume.

Car Financing -- ILAP
Developed and used with Department of Social Sciences (Economics) in Spring 1994. MAJ Roune (Econ) developed video for Fall 1994. COMAP published this ILAP in NSF Sample Book.

D. B. Cooper -- In-class activity
Developed by John Dossey (Visiting Professor at West Point from Illinois State) and used in Fall 1993 as a course-end activity involving discrete dynamical systems and differential calculus.

Unemployment -- In-class activity
Developed by Jack Picciuto & Paul Laumakis and used in Fall 1994 as an activity involving discrete dynamical systems and calculus.

Analysis of Military Retirement Pay -- Project
Developed by Dave Thiede and used in Fall 1995. Not an ILAP, but could be developed as one with Economics.

Airborne Personnel (personnel turnover in airborne divisions) -- Project
Developed by Mike Huber and Doug Bentley and used in Fall 1995. Not an ILAP, but could be developed as one with History.

Making Water in Space -- ILAP
Developed by Randy Rotte and Department of Chemistry and used in Spring and Fall 1996.

MA104 -- Single Variable Calculus and Differential Equations

Flying Strategies -- ILAP

Developed and used with CME in Spring 1993. Kip Nygren (CME) delivered lecture in an auditorium. Video developed afterward.

Bungee Cord Jumping -- ILAP

Developed and used with CME in Spring 1993. Steve Ressler delivered lecture on connecting bungee jumping to the trebuchet in an auditorium. Video developed afterward. COMAP published in the NSF Sample Book in February 1995. Also a video was developed with CME.

Aerobic Capacity -- ILAP

Developed by Shep Barge and Todd Crowder (Department of Physical Education) and used in Spring 1994. Video developed afterwards. Presented by Dave Jensen at MAA/AMS summer meetings in August 1994 at Minneapolis, MN. Now entitled "Getting Fit with Mathematics" and included in this MAA volume.

Vibration of an Airplane Wing -- ILAP

Developed and used with Guy Harris (CME) in Spring 1994. Lecture presented in an auditorium. Video taken of lecture. Now entitled "Flying with Differential Equations" and included in this MAA volume.

Clinic Profit Management -- ILAP

Developed by Joe Schulz & Dean Dudley, Department of Social Sciences, and used in Spring 1995.

Clinic Profit II -- ILAP

Developed by Joe Schulz & Dean Dudley, Department of Social Sciences, and used in Spring 1995.

Wheel Suspension Design -- ILAP

Developed by Joe Schulz & Bob Schulz (CME) and used in Spring 1995. Lecture given in an auditorium by Bob Schulz. Video taken of lecture.

MA104 -- Single Variable Calculus and Differential Eqns (continued)

Bass Population -- Project
Developed by Jeff Strickland, integrates discrete dynamical systems and differential calculus used as Project 1 in Spring 1992.

Terrain Analysis -- Project
Developed by Mike Talbott and used in Spring 1992. An optimization problem.

Telemetry Data Interpretation -- Project
Developed by Jack Robertson and used in Spring 1991. Adopted and refined by Mike Talbott and used in Spring 1992. A numerical integration problem. Later Jeff Strickland developed as missile testing problem.

Real Estate Taxation -- Project
Developed by Mike Jaye from Ithaca College Calculus Project and used as a project in Fall 1992. An optimization problem.

Centers of Mass -- Project
Developed by Shep Barge from Project CALC materials and used in Fall 1993.

Road Construction -- Project
Developed by Dennis Polaski from an actual military experience in Honduras 1984-85. Used in Spring 1994. An optimization problem with multivariate aspects.

Light Refraction -- Project
Developed by Shep Barge from UMAP Module #341. Used in Spring of 1994.

Airport Construction -- Project
Developed by Joe Schulz and Dennis Polaski and used in the Fall of 1994. An integration problem.

Forest Fire Fighting -- ILAP
Developed by Dennis Polaski, Doug Bentley, and Bill Doe (GEnE) and used in Spring 1996. Video introduction and expert briefing was prepared and used.

MA104 -- Single Variable Calculus and Differential Eqns (continued)

Water Reservoir Management -- ILAP

Developed by Dennis Polaski, Archie Wilmer, and Brian Dosa (CME) and used in Spring 1996. Video introduction was prepared. Problem uses integration, related rates, and optimization.

Parachute Jumping -- ILAP

Developed by Dennis Polaski, Billy Tollison, and James Nichols (Physics) and used in Spring 1996. Video introduction, solution, and extension were prepared. Problem solves a falling-body problem with differential equation models.

MA205 Calculus II -- Multivariable Calculus

Vehicle Collision -- ILAP

Developed by Dennis Polaski with T.J. Creamer, Physics, and used in Fall 1995. Published by COMAP as part of Project INTERMATH Sample Book to NSF.

Missile Trajectory -- Project

Developed by Dave Olwell and used in Fall 1991. Used with a field artillery scenario in Fall 1993 and Fall 1994.

Cobb-Douglas Production Model -- ILAP

Developed by Dave Olwell with Tom Daula, Social Sciences Department, and used in Fall 1991. Used again in Fall 1992 with a different scenario. Used with a recruiting command scenario in Fall 1994.

Tank Suspension -- Project

Developed by Dave Olwell and used in Fall 1991. Spring-mass problem. This is a precursor to the wheel suspension problem used in MA104 in Spring 1995.

Helicopter Flight Planning -- Project

Used in Fall 1992.

SIPE (Soldier Integrated Protective Ensemble) -- Project

Multivariate optimization problem used in Fall 1993.

MA206 Probability & Statistics

Great Lakes Pollution -- Carry-through ILAP

Developed by Bill Fox and Bill Goetz and used in Spring 1995 as three projects in 3 parts. Linked to previous ILAP in MA103 and in-class activity in MA104. Included in NSF Sample Book as carry-through ILAP published by COMAP.

Combat Vehicle Identification System (CVIS) -- Project

Reliability project used in Spring 1994.

Vehicle Accident Analysis -- Project

Distributions project used in Spring 1994 as four parts.

Remotely Piloted Vehicle -- Project

Reliability project used in Fall 1993.

Dinosaur Exhibition Park -- Project

Distributions project used in Fall 1993 as two parts.

Long Range Surveillance Mission Analysis -- Project

Reliability project used in Fall 1995. Informally coordinated with the Department of Military Instruction.

Tactical Operations Center Information Management -- Project

Information systems modeling project used in Fall 1995.

Cadet Database Analysis -- Project

Exploratory data analysis on cadet performance. In phase one, descriptive statistics are used to examine selected random variables. In phase two, inferential statistics are used to estimate future behavior and test hypotheses about cadet performance. Used in Spring 1996.

MA364 Engineering Mathematics

Trajectory Deviation of a Cruise Missile -- ILAP

Draft written by Peter Plostins, Army Research Laboratory, in Fall 1994.

MA366 Vector Calculus and Intro to PDEs

Aquifer Contamination -- ILAP

Developed by Dick Jardine, Mike Jaye, and Stan Thomas (GEnE) and used in Spring 1995 & 96. Project is used in preparation for EV394 Hydrogeology. This ILAP is contained in this MAA volume.

Flow Nets -- ILAP

Developed by Dick Jardine and Chris King (GEnE) and used in 1996.

COMAP Sample Book

COMAP published four ILAPs in 1995 in a sample book entitled *An Introduction to the Interdisciplinary Lively Applications Project: Four Sample Projects.* Included in this volume are the following ILAPs: "Car Financing," "Bungee Cord Jumping," "Vehicular Collision," and "Lake Pollution." These four ILAPs and many others being written through Project INTERMATH will be available from COMAP.